Aircraft Surveillance Systems

The Communication, Navigation and Surveillance (CNS) systems provide air traffic controllers with the information necessary to ensure the specified separation between aircraft and efficient management of airspace, as well as assistance to flight crew for safe navigation. However, the radar systems that support air traffic management (ATM), and in particular air traffic control (ATC), are at their operational limit. This is particularly acute in the provision of the ATC services in low altitude, remote and oceanic areas. Limitations in the current surveillance systems include unavailability of services in oceanic and remote areas, limited services during extreme weather conditions and outdated equipment with limited availability of spare parts to support system operation. These limitations have resulted in fatal accidents.

This book addresses the limitations of radar to support ATC in various operational environments, identified and verified by analysing five years of safety data from Avinor, the Air Navigation Service Provider (ANSP) in Norway. It derives a set of taxonomy and from this develops a causal model for incident/accident due to limitations in the surveillance system. The taxonomy provides a new method for ANSPs to categorize incidents while the causal model is useful for incident/accident investigations. The book also provides theoretical justifications for the use of Automatic Dependent Surveillance Broadcast (ADS-B) to overcome the limitations of radar systems and identify areas of improvement to enable seamless ATC services.

Written in a style that makes it accessible to non-specialists, *Aircraft Surveillance Systems* will be of interest to many in the field of aviation, particularly ATM, safety and accident/incident investigation. It will also offer a useful reference on this vital topic for air traffic management courses.

Dr Busyairah Syd Ali is Senior Lecturer at the Department of Software Engineering at University of Malaya, Malaysia. Strong passion and an interest in aviation led her to undertake a PhD in Air Traffic Management (ATM), which she completed at the Centre for Transport Studies, Imperial College London, United Kingdom. Dr Syd Ali has investigated the limitations of radar that lead to aircraft incidents, and has developed a safety assessment framework for a new surveillance technology called Automatic Dependent Surveillance Broadcast (ADS-B), and assessed its impact on air traffic management operations. She has outstanding expertise (15 years in aviation), gained through studies in Air Traffic Management as well as her work experience with the Department of Civil Aviation Malaysia and, throughout the course of her PhD, exposure to the operations of NATS UK, Avinor Norway, EUROCONTROL and easyJet.

Aircraft Surveillance Systems

Radar Limitations and the Advent of the
Automatic Dependent Surveillance
Broadcast

Busyairah Syd Ali

Routledge
Taylor & Francis Group

LONDON AND NEW YORK

First published 2018
by Routledge

2 Park Square, Milton Park, Abingdon, Oxfordshire OX14 4RN
52 Vanderbilt Avenue, New York, NY 10017

Routledge is an imprint of the Taylor & Francis Group, an informa business

First issued in paperback 2019

British Library Cataloguing in Publication Data
A catalogue record for this book is available from the British Library

Library of Congress Cataloging in Publication Data
A catalog record for this book has been requested

ISBN: 978-1-4724-7797-2 (hbk)
ISBN: 978-0-367-88171-9 (pbk)

Typeset in Bembo
by Wearset Ltd, Boldon, Tyne and Wear

Contents

Figures

Tables

Abbreviations

ADS   Automatic Dependent Surveillance
ADS–B   Automatic Dependent Surveillance Broadcast
AMSS   Aeronautical Mobile Satellite Service
ANSP   Air Navigation Service Provider
ASAS   Airborne Separation Assistance System
ASDE   Airport Surface Detection Equipment
ATC   Air Traffic Control
ATCo   Air Traffic Controller
ATM   Air Traffic Management
ATN   Aeronautical Telecommunication Network
ATS   Air Traffic Services
CAA   Civil Aviation Authority
CDTI   Cockpit Display of Traffic Information
CNS   Communication, Navigation and Surveillance
CNS/ATM   Communication, Navigation, Surveillance and Air
            Traffic Management
DL   Data Link
DME   Distance Measuring Equipment
EEC   EUROCONTROL Experimental Centre
FAA   Federal Aviation Administration
FIR   Flight Information Region
FIS–B   Flight Information Service Broadcast
FMECA   Failure Mode Effects and Criticality Analysis
FOM   Figure of Merit
FSPN   Fluid Stochastic Petri Nets
FTA   Fault Tree Analysis
GBAS   Ground Based Augmentation System
GBT   Ground Based Transceiver
GNSS   Global Navigation Satellite System

GPS	Global Positioning System
HF	High Frequency
HFDL	High Frequency Data Link
HFOM	Horizontal Figure of Merit
HPL	Horizontal Protection Level
IATA	International Air Transportation Association
ICAO	International Civil Aviation Organization
IFR	Instrument Flight Region
ILS	Instrument Landing System
INS	Inertial Navigation System
IRS	Inertial Reference System
ISO	International Organization for Standardization
ITP	In-Trail Procedure
MLAT	Multilateration System
MSSR	Monopulse Secondary Surveillance Radar
NAC	Navigational Accuracy Category
NIC	Navigational Integrity Category
NUC	Navigational Uncertainty Category
OEM	Original Equipment Manufacturer
OSI	Open System Interconnection
PBN	Performance Based Navigation
PSA	Probabilistic Safety Assessment
PSR	Primary Surveillance Radar
QoS	Quality of Services
RAIM	Receiver Autonomous Integrity Monitoring
RCP	Required Communication Performance
RNAV	Area Navigation
RNP	Required Navigation Performance
RSP	Required Surveillance Performance
RTCA	Radio Technical Committee for Aeronautics
SADT	System Analyses and Design Technique
SBAS	Satellite Based Augmentation System
SIL	Source Integrity Level
SIS	Signal In Space
SL	Safety Level
SMR	Surface Movement Radar
SSR	Secondary Surveillance Radar
TDOA	Time Difference of Arrival
TIS-B	Traffic Information Service Broadcast
TSO	Technical Standard Orders
UAT	Universal Access Transceiver

UAV	Unmanned Aerial Vehicle
VDL	VHF Data Link
VFR	Visual Flight Region
VHF	Very High Frequency
VOR	VHF Omnidirectional Radio Range
WAAS	Wide Area Augmentation System
WAM	Wide Area Multilateration
WGS	World Geodetic System

1 Communication, Navigation, Surveillance and Air Traffic Management (CNS/ATM)

1.1 Background

Conventional air navigation systems such as radars, Instrument Landing System (ILS), VHF Omnidirectional Radio Range and Distance Measuring Equipment (VOR/DME), used for airspace surveillance, navigation and communication are ground-based systems. However, these systems suffer from a number of drawbacks including accuracy limits, range and line-of-sight limitations, being site-critical, the requirement for many installations and the considerable expense required for acquisition and maintenance. While significant advances have been made in hardware and software, the technology principle employed is typically more than 60 years old. Furthermore, these systems are unable to evolve to meet increasing traffic demands around airports, and are difficult to implement over large parts of the earth for example, because of remoteness and inhospitable terrain.

In 1983, the International Civil Aviation Organization (ICAO) gave the task of studying, identifying and assessing new concepts and technologies in the field of air navigation, including satellite technology, to a special committee. The Future Air Navigation Systems (FANS) Committee gathered together aviation specialists from around the world. In such a global forum, these specialists developed the blueprint for the system that would meet the needs of the aviation community well into the next millennium (ICAO, 1998a). The FANS concept, which came to be known as the Communication, Navigation, Surveillance and Air Traffic Management (CNS/ATM) system, involves a complex and interrelated set of technologies, largely dependent on satellites, in order to overcome certain limitations of the existing systems. By adopting an approach whereby satellites would play a major role in communication, navigation and surveillance, the FANS Committee determined that states could substantially increase signal coverage over large parts of the earth with fewer infrastructures.

ICAO in 1992 endorsed CNS/ATM as the sole air navigation services (ANS) system for global application (ICAO, 1998b).

1.2 Communication, Navigation, Surveillance and Air Traffic Management (CNS/ATM)

ICAO defined CNS/ATM as "Communication, Navigation and Surveillance systems, employing digital technologies, including satellite systems together with various levels of automation, applied in support of a seamless global air traffic management system" (ICAO, 2000). The aim of CNS/ATM is to develop a comprehensive and unified system to support the provision of Air Traffic Services (ATS) to meet growth in air travel demand with associated improvements in safety, efficiency and regularity of air traffic, providing the desired routes to the airspace users, and homogenizing the use of equipment in different regions. CNS/ATM is underpinned by a high level of automation which reduces the dependency on the human and eliminates the current constraints to optimize the airspace. The distinct features of CNS/ATM are (ICAO, 2000):

- mix of satellite and ground-based systems – which enable internetworking for data transfer of communication, navigation and surveillance systems from technical sites to operational units to provide complete situational awareness to controllers and pilots;
- global coverage – which enables complete ATC services despite the geographical structure obstacles;
- seamlessness – whereby continuous and reliable services are available without fail to ensure safety;
- interoperable systems – whereby the system is designed as redundant architecture to provide uninterrupted services;
- use of air-ground data link – which enables synchronized situational awareness to controllers and pilots;
- use of digital technologies – to mitigate the limitations of analogue technologies such as noise interruption and to adapt to new digital application systems;
- various levels of automation – whereby more computer applications are used to aid controllers and pilots to perform the various job functions.

Figure 1.1 depicts the paradigm shift in ATM technologies, from the current CNS systems to the new CNS/ATM systems that are a mix of

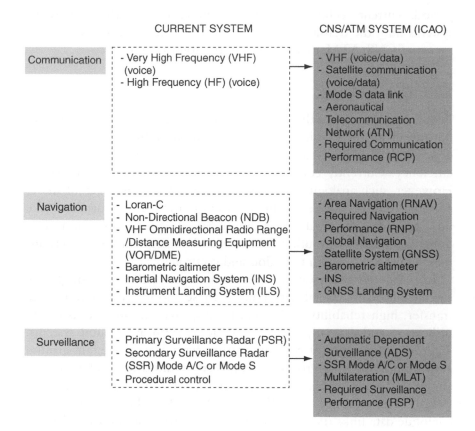

Figure 1.1 Paradigm shift in ATM technologies.

Source: modified from Vismari and Camargo, 2005.

satellite technology and the best of the line-of-sight systems. The new technologies have the potential to support advanced ATM applications such as Cockpit Display of Traffic Information (CDTI) (ICAO, 2003a) that provides situational awareness to pilots and In-Trail Procedure (ITP) (EUROCONTROL, 2009a) to give an aircraft more flexibility for efficient navigation especially in oceanic en-route areas. This in return benefits the airlines in terms of fuel consumption and most importantly reduces the environmental effects (Federal Aviation Administration, 2012). New supported applications are discussed in Chapter 4.

ICAO has developed a Global CNS/ATM Plan (ICAO, 2002a). Contracting states are to develop and implement a National CNS/ ATM Plan (ICAO, 2000) based on the ICAO Global Plan. For a period, current technology systems will co-exist with CNS/ATM systems until the transition to CNS/ATM is complete. The main elements of CNS/ATM are addressed in the following sections.

1.2.1 Communication

People and systems on the ground must communicate with the aircraft during all phases of flight. Good communications with timely and dependable availability are the cornerstone of operational safety and efficiency. Currently communication is primarily by means of voice. However, such analogue transmissions suffer from a number of shortcomings: they do not permit high rates of transmission of data and take up a great deal of valuable and diminishing frequency spectrum. This limits automation of routine functions and consequently the decision-making process for both the pilots and controllers.

In CNS/ATM systems, communications will increasingly be carried out using digital data links as these allow a high rate of data transfer, high reliability and integrity, improved frequency spectrum utilization and crucially, better interfacing with automated systems. There are two types of communication systems in place; air-ground communication and ground-ground communication. The current air-ground communication system relies on Very High Frequency (VHF), High Frequency (HF) and Ultra-High Frequency (UHF) analogue data links (radio frequency) for en-route and terminal areas, and Aeronautical Mobile Satellite Service (AMSS) for oceanic and remote continental airspace (ICAO, 2000), while the ground-ground communication relies on VHF data link. According to Hansman (1997), the current flight procedures and route structures have been developed and named based on the voice communication capabilities over low bandwidth VHF and HF links, resulting in limited coverage. Figure 1.1 shows the evolution of the communication technologies.

Future communication systems are based on digital data links such as High Frequency Data Link (HF DL), VHF Data Link Mode 4 (VDL-Mode 4), Mode S Extended Squitter and Universal Access Transceiver (UAT). Data link technologies enable uplink and downlink of four dimensional (4D) waypoints (latitude, longitude, altitude, time) and other data to pilots and controllers. Controller-Pilot Data Link Communication (CPDLC) is an example of a data link application that relies on HF DL, VDL and satellite communication

(SATCOM). The implementation of the digital data links has the potential to change the communication of control instructions in the event of analogue voice link failure (Hansman, 1997).

Moreover, the analogue voice communication is prone to many limitations to the users, e.g. limited coverage, accessibility, capability, integrity and security. The voice communication performance, based on the radio frequency can reduce due to interference issues, frequency congestion and noise. This can happen, even though there are specific aviation frequency bands allocated for the ATC use. In addition, due to the different accents of the pilots and controllers, voice communication can lead to misinterpretation of information, which may cause undesired incidents. Despite its limitations, voice communication via radio frequency is still the main mode of communication between pilots and controllers in the ATC environment. Voice communication channels are regarded essential for ATC, since they act as a backup during the worst case (unavailability of surveillance and navigation functions) to enable continuous air traffic services to the users.

The implementation of enhanced modes of data link is envisioned to overcome the limitations discussed above. Therefore, the need for reliable digital data link technologies is crucial. However, the new digital communication technologies have to comply with the Required Communication Performance (RCP) (ICAO, 2006c) set by ICAO. The future communication systems in ATM are envisioned to be a mix of voice and data communication via high-speed digital data links.

1.2.1.1 Aeronautical Telecommunication Network (ATN)

The first step in implementing CNS/ATM, is the establishment of an efficient networking system for the communication of different forms of data, including text, radar, graphics and voice. This requires the use of a combination of terrestrial and satellite-based systems. The current system, the Aeronautical Fixed Telecommunication Network (AFTN), does not have the capability to support the future data requirements of CNS/ATM (ICAO, 2000). Therefore, ICAO proposes the use of the Aeronautical Telecommunication Network (ATN) that comprises application entities and communication services. These make the ground elements, air-ground networks and airborne data networks interact via the International Organization for Standardization (ISO) Open System Interconnection (OSI) reference model based protocol and services interface (ICAO, 1999a). ICAO has standardized the following data links in the context of ATN (ICAO, 1999c):

- Aeronautical Mobile Satellite Service (AMSS), using satellites for communication, both geostationary and non-geostationary satellites, allowing communication by voice and data on a global range.
- VHF Data Link (VDL), using techniques of data communication in VHF bands. They are of types Mode 2, Mode 3 and Mode 4 with differentiation by their characteristics of modulation, control for access to the physical environment and, especially, data transfer rates.
- Mode S Extended Squitter (ES), operating on 1090/1030 MHz to communicate data in a bidirectional manner between air and ground elements with nominal rates of 4 Mbits/s (uplink) and 1 Mbits/s (downlink) (ICAO, 1998b).
- Universal Access Transceiver (UAT), a broadcast data link operating on 978 MHz, with a modulation rate of 1.041667 Mbps (ICAO, 2009).
- HF Data Link (HFDL), which is the union between the characteristics of long-range electromagnetic propagation in the HF spectrum and digital data modulation, providing data communication in remote areas.

Vismari (2007) illustrates the CNS/ATM communication environment based on the ATN in Figure 1.2. ICAO categorized the application entities (AE), which are the functionalities of the ATN used by end systems (ES) in the air traffic system, into air-ground application entity and ground-ground application entity (ICAO, 1999a). The air-ground AE enables communication between ES on the ground (ATS units) and ES in the air (aircraft). Examples of applications in this category include:

- Automatic Dependent Surveillance Broadcast (ADS-B), which provides the aircraft position and other important information to the ES;
- Controller-Pilot Data Link Communication (CPDLC), which provides the ability to establish a peer-to-peer message communication between pilots and controllers;
- Flight Information Services (FIS), which allows pilots to request and receive flight information services; and
- Traffic Information Service Broadcast (TIS-B), which transmits radar surveillance information from the ground to the aircraft in the air.

The ground-ground AE allows communication between ES on the ground (ATS units). The AE in this category are the ATS Message

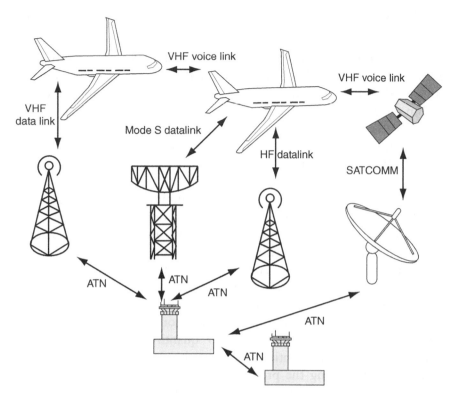

Figure 1.2 CNS/ATM communication environment.
Source: modified from Vismari, 2007.

Handling Service (ATSMHS), enabling the exchange of messages between ATS end users; and the Inter-Communication Centre (ICC), which provides message communication between ATS centres for notification, coordination and transfer of control activities.

An application developed based on the ATN is SESAR's and Next-Gen's System Wide Information Management (SWIM) (SESAR Joint Undertaking, 2011). SWIM is a holistic approach enabling information sharing including flight information, weather, aeronautical information and surveillance information among the stakeholders and the airspace users using a secure and flexible system (an intranet). SWIM infrastructure, over which the data are distributed, is interoperable (ground/ ground and air/ground). Its data communication link may differ from one user to another depending on available facilities. For SESAR's SWIM, the PAN European Network System (PENS) will provide the ground/ground data link.

display error and flight technical error (ICAO, 1999b). It refers to the level of accuracy required for a given block of airspace and/or a specific instrument procedure, e.g. a RNP of 10 means that a navigation system must be able to calculate its position to within a circle with a radius of 10 nautical miles. The level of RNP that an aircraft is capable of determines the separation required between it and other aircraft. Hence, the RNP values have to be more stringent for dense airspace, around noise sensitive areas or terrain areas compared to oceanic airspace.

The RTCA extended the RNP definition to include integrity, continuity and availability (RTCA, 1998). It was then known as Required Navigation Performance for Area Navigation (RNP-RNAV). The difference between RNP and area navigation (RNAV) is that the RNAV is a navigation specification that does not include a requirement for on-board navigation performance monitoring and alerting.

In order to implement a more practical navigation specification, ICAO developed Performance Based Navigation (PBN). PBN specifies that the aircraft RNP and RNAV system performance requirements are to be defined in terms of accuracy, integrity, continuity, availability and the functionalities required to operate in particular airspace supported by the appropriate navigation infrastructure (ICAO, 2008). The performance requirements are identified in the navigation specification, which also states the choice of navigation sensor and equipment that may be used to meet the performance requirements. PBN requirements depend on the ATC environment, communication, surveillance, navigational aids infrastructure, non-RNAV means of navigation available, functional and operational capabilities required to meet the Air Traffic Management application and the degree of redundancy required to ensure continuity of operations. The PBN provides specific implementation guidance in order to facilitate global harmonization. In this book, the term "RNP parameters" refers to accuracy, integrity, continuity and availability, which are defined in Chapter 4.

1.2.3 Surveillance

Surveillance, which refers to the methods used for keeping track of aircraft, is the third element of CNS/ATM. The surveillance function implementation includes sensors, display system and operational procedures, and provides air traffic controllers with the position of aircraft in order to perform separation management and to effectively manage a given airspace. Depending on the type of surveillance sensor, additional information such as aircraft identification and velocity is also presented. Furthermore, the surveillance function supports a number of other applications such as trajectory prediction, conflict detection and situational awareness.

Requirements for an Air Traffic Control (ATC) surveillance system depend on the applications. However, no single surveillance system is capable of meeting the surveillance requirements for all phases of flight in all types of airspace with traffic conditions that vary significantly from low to high-density traffic terminal areas. The current surveillance system in use consists of: Primary Surveillance Radar (PSR), Secondary Surveillance Radar (SSR), Monopulse Secondary Surveillance Radar (MSSR), Surface Movement Radar (SMR) and Multilateration (MLAT) systems. These technologies are explained in detail in Chapter 2. The evolution of the surveillance systems is illustrated in Figure 1.1. Recently a new surveillance technology, called Automatic Dependent Surveillance (ADS), has emerged, and is envisioned to support many new surveillance applications to meet future air traffic forecasts. ADS exploits the navigation and communication functions. The availability of different types of surveillance technologies provides flexibility to choose the most affordable and effective surveillance system suitable for the required operations, based on the operational environment. However, in order to maintain harmonization of the surveillance function, all the operational requirements have to be translated into a series of surveillance performance parameters irrespective of the surveillance technology.

1.2.3.1 Required Surveillance Performance (RSP)

The Required Surveillance Performance (RSP) is a set of well-quantified surveillance performance requirements such as capacity, availability, accuracy and update rate. Any single or combination of surveillance systems meeting the targets set for the parameters is considered operationally acceptable (ICAO, 2000). The only RSP document available to date is known as the Blue Book (EUROCONTROL, 1997), which is specifically meant for Primary Surveillance Radar (PSR) and classical Secondary Surveillance Radar (SSR Mode A/C). Therefore, it is not applicable to any new surveillance technology performance requirements. With the emergence of new surveillance technologies, all surveillance systems in the European Union are legally obliged to comply with the Single European Sky Essential Requirement (ER), which states:

- Surveillance systems shall be designed, built, maintained and operated using appropriate and validated procedures in such a way as to provide the required performance applicable in a given environment (surface, TMA, en-route) with known traffic characteristics and exploited under an agreed and validated operational concept, in particular in terms of accuracy, coverage, range and quality of service.
- The surveillance network within the European Air Traffic Management Network (EATMN) shall be such as to meet the requirements of accuracy, timeliness, coverage and redundancy. The surveillance network shall enable surveillance data to be shared in order to enhance operations throughout the EATMN.

(EUROCONTROL, 2008)

This high-level requirement will be augmented by an Implementing Rule (IR), the Surveillance Performance and Interoperability Implementing Rule (SPI-IR) (EUROCONTROL, 2011a), which specifies how the essential rule is to be achieved. The SPI-IR is a legal requirement and includes regulations and general surveillance performance requirements (explained in Chapter 4). This implementing rule will remain in place until a generic global RSP is mandated by ICAO. This book adopts the surveillance performance requirements stipulated in this implementing rule.

1.2.4 Air Traffic Management (ATM)

Air Traffic Management (ATM) is a broadly defined function that includes air traffic services, air traffic flow management and airspace management. Its objective is to keep aircraft separated and enable aircraft operators to meet their planned times of arrival and departure while adhering to preferred flight (ICAO, 2002a). Integration of the new CNS technologies into the ATM will enable Air Traffic Services (ATS) providers to improve efficiency. By being better able to both accommodate an aircraft's preferred flight profile and reduce the minimum separation, aircraft operators and service providers could achieve reduced operating costs and minimize delays, while simultaneously freeing up additional airspace and increasing capacity. Figure 1.3 summarizes the benefits of the new CNS systems to the ATM.

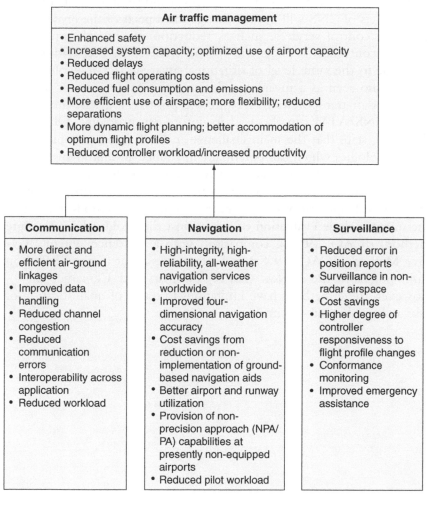

Figure 1.3 A high-level view of benefits of the new CNS system to ATM.
Source: ICAO, 2002a.

1.3 Impacts of evolution in CNS

The paradigm shift in CNS, outlined in Figure 1.1, due to developments in technology, directly impacts the characteristics of ATM. Such impacts include the ATM system's ability to effectively manage the separation between aircraft due to: an increase in the accuracy, integrity and reliability of surveillance data, the larger coverage area, a reduction of congestion in the communication channels and the potential for improvements in the detection and resolution of conflicts. The impacts of CNS will in turn change the aspects of the provision of air traffic control services, such as reduction in control instructions from the controllers as pilots will also be involved in a separation situation due to the same level of situational awareness. The evolution in CNS is also seen as a means to eliminate the constraints on the safe growth of air transportation. Despite the positive impacts, the evolution of CNS/ATM can also lead to unforeseen problems. Oliveira *et al.* (2009), state that the main challenge in CNS is the complexity of the technologies which can potentially introduce an unknown number of new failure modes. Therefore, it is crucial to assess and validate the safety of these technologies prior to their implementation to support safety-critical applications. The current relevant methods for safety assessment include evaluation of risk against threshold value (Absolute Method) (ICAO, 1998c), comparison with a reference system (Relative Method) (ICAO, 1998c) and safety assessment method based on Fluid Stochastic Petri Nets (FSPN) (Vismari and Camargo, 2008). However, these methods have limitations in terms of quantification of risks and are thus sub-optimal.

2 Air Traffic Control (ATC) surveillance environment

2.1 The need for ATC surveillance

Surveillance acts as the "eyes" of Air Traffic Control (ATC). The capability to accurately and reliably determine the position of an aircraft at a specific time has a direct influence on the separation distances required between aircraft (i.e. separation standards), and therefore, on how efficiently a given airspace may be utilized.

In areas without surveillance coverage, where ATC is reliant on pilots to verbally report their position, aircraft have to be separated by relatively large distances to account for the uncertainty in the estimated position of aircraft and the timeliness of the information. Separation requirements between aircraft are discussed further at the end of the chapter. Conversely in terminal areas where accurate and reliable surveillance systems are available and aircraft positions are updated more frequently, the airspace can be used more efficiently to safely accommodate a higher density of aircraft. It also allows aircraft vectoring for efficiency, capacity and safety purposes.

ATC surveillance serves to close the gap between ATC expectations of aircraft movements based on clearances or instructions issued to pilots, and the actual trajectories of the aircraft (ICAO, 2007c). In this way, it indicates to ATC when expectations are not matched, providing an important safety function. Surveillance therefore provides "blunder" (false position) detection.

The demand for increased flexibility to airspace users by reducing restrictions associated with flying along fixed routes requires high performance (accuracy, integrity, continuity, reliability) navigation capability on board the aircraft. Equally, accurate surveillance is required to assist in the detection and resolution of any potential conflicts associated with the flexible use of the airspace, which is likely to result in a more dynamic environment. This concept will be applied to enable e.g. In-Trail Procedure (ITP) (EUROCONTROL, 2009a), which allows a pilot to navigate flexibly in en-route and oceanic areas by having complete situational awareness.

Surveillance is required to support automated alerting systems such as Short Term Conflict Alert (STCA) for the ATC function. Automated alerting systems are based on the principle that the ability to accurately track aircraft enables ATC to be alerted when an aircraft is detected to:

1   deviate from its assigned altitude or route, or
2   the predicted future positions of two or more aircraft conflict.

Surveillance also supports Minimum Safe Altitude Warnings (MSAW), danger area warnings and other similar alerts (ICAO, 2007c).

Finally, surveillance is also used to update flight plans and improve estimates of future waypoints, thereby reducing the workload for pilots in providing voice reports on reaching waypoints to the ATC. Therefore, surveillance is a crucial element of air traffic control.

2.2 Current ATC surveillance

Conventional ATC Surveillance is based on voice position reporting by the pilot to the controllers on the ground via radio communication. While manoeuvring in an oceanic area, the pilot has to report every 20–30 minutes to the controller within the control area via VHF or HF frequency, and then the controller has to repeat it for verification purposes.

A generic ATC surveillance system includes sensors (technology), communication links, a surveillance data distribution system, a surveillance data processing system, a surveillance data analysis tool, display, surveillance applications and users. Such a generic system is illustrated in Figure 2.1.

The main components in Figure 2.1 are:

- Surveillance Data Distribution System – converts the data into a standardized format (e.g. ASTERIX) and then transmits the data to other equipment;
- Surveillance Data Processing System – extrapolates plots to generate track state vector;
- Surveillance Data Analysis Tool – analyses data performance;
- Safety Net – tools meant to prevent imminent or actual hazardous situations from developing into major incidents or even accidents;

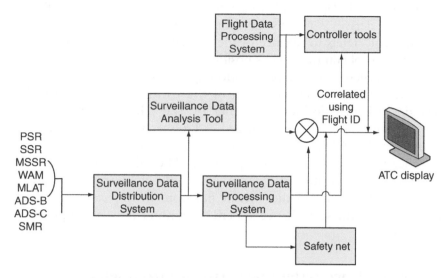

Figure 2.1 Generic ATC surveillance system.

- Flight Data Processing System – stores, displays and updates flight plan.

Currently, surveillance systems provide controllers with the surveillance picture for their control area (sector). In the future, pilots will also have a similar surveillance picture on-board. Therefore, this advancement will impact the current ATC operations.

2.3 Surveillance sensor categories

Surveillance sensor technologies can be placed in three categories: Non-Cooperative Independent, Cooperative Independent and Cooperative Dependent.

The term "Non-Cooperative" refers to the ability to detect the position of a target without relying on the response from the target to the transmitted signal by the sensor. On the other hand, the term "Cooperative" refers to reliance of the sensor on the target's reply to the transmitted signal (interrogation) to derive the target position. A piece of equipment (i.e. a transponder) attached to the target responds to the sensor interrogation. The transponder is a radio signal receiver and transmitter that receives at a frequency 1030 MHz and transmits on 1090 MHz.

The term "Independent" refers to the ability of the surveillance system to derive a target's position on its own, while the term "Dependent" refers to reliance of the surveillance system on an external system to obtain the target's position, e.g. dependency on a navigation system such as the Global Positioning System (GPS).

Table 2.1 shows the categories of the existing surveillance technologies. Manual position reporting by the pilot to the ATC via radio communication is categorized as Cooperative Dependent, due to the fact that ATC on the ground is required to respond to the call in order to report the aircraft's position. The pilot is dependent on the on-board navigation systems such as GPS or Inertial Navigation System (INS) to obtain the aircraft position. In this case the pilot/aircraft acts as the surveillance system to transmit the aircraft position to the ATC.

2.4 Surveillance system performance parameter and characteristics

A surveillance system may be characterized in terms of the following parameters (ICAO, 2007c):

- coverage volume – the volume of airspace in which the system operates to specification;
- accuracy – a measure of the difference between the estimated and true position of an aircraft;

Table 2.1 Categories of existing surveillance technologies

Surveillance category	Surveillance technology
Non-Cooperative Independent	• Primary Surveillance Radar (PSR)
Cooperative Independent	• Secondary Surveillance Radar (SSR) Mode A/C • Secondary Surveillance Radar (SSR) Mode S • Monopulse Secondary Surveillance Radar (MSSR) • Multilateration (MLAT)
Cooperative Dependent	• Manual Position Reporting (voice) • Automatic Dependent Surveillance Contract (ADS-C) • Automatic Dependent Surveillance Broadcast (ADS-B) IN/OUT

- integrity — an indication that the aircraft's estimated position is within a stated containment volume of its true position. Integrity includes the concept of an alarm being generated if this ceases to be the case, within a defined time to alarm. Integrity can be used to indicate whether the system is operating normally;
- update rate — the rate at which the aircraft's position is updated to users;
- reliability — the probability that the system will continue operating to specification within a defined period. This is also called continuity;
- availability — the percentage of the total operating time during which the system is performing to specification.

The parameter values may differ for the various surveillance applications supported by the surveillance system during the different phases of flight. For example, more stringent values for the update rate and accuracy are required to enable enhanced separation in terminal areas compared to en-route oceanic areas. The performance characteristics of each surveillance sensor discussed in the previous section is given in Table 2.2.

The same performance parameters can be used to assess different surveillance technologies. However, when applied to different technologies, the definition of the parameters may change slightly. The performance parameters for the ADS-B system are defined in Chapter 4.

2.5 Required surveillance applications

Surveillance technologies are meant to make possible various ground and airborne surveillance applications (RTCA, 2002) to support air traffic control, airspace management and aircraft navigation in en-route, terminal and airport surface areas. Surveillance data from different surveillance technologies have different levels of quality and performance (i.e. accuracy, integrity, update rate). The capabilities of the surveillance sources are used as a baseline to develop surveillance tools to support the ground and airborne surveillance applications required. The surveillance applications in three categories are shown in Table 2.3.

Another essential surveillance application to aid pilots and controllers to prevent imminent or actual hazardous situations from developing into major incidents or even accidents is called the safety net (shown in Figure 2.1 as part of the ATC surveillance system). Safety

Table 2.2 Surveillance sensor performance characteristics

Surveillance technology	Coverage	Accuracy	Integrity	Update period (seconds)
Primary Surveillance Radar (PSR)	S-band 60–80 NM L-band 160–220 NM	In range: 0.1 NM RMS or 0.2 NM 2σ In azimuth: 0.15 degrees RMS or 0.3 degrees 2σ	No integrity report provided	4–15
Secondary Surveillance Radar (SSR) (Mode A/C)	200 NM–250 NM	In range: 0.03 NM RMS In azimuth: 0.07 degrees RMS or 0.14 degrees 2σ for random errors	No integrity report provided	4–15
Secondary Surveillance Radar (SSR) (Mode S)	200 NM–250 NM	Same as SSR (Mode A/C)	No integrity report provided	4–12
Multilateration (MLAT)	200 NM	10–500 metres	No integrity report provided	1–5
ADS-B	200 NM–250 NM	Determined by the aircraft avionics and independent of range from sensor. For GPS, 95% less than 0.1 NM	Position integrity guaranteed to 1×10^{-7} due to RAIM algorithm in avionics. Integrity value is downlinked in the ADS-B message	0.5–2

Source: ICAO, 2007d.

Table 2.3 Surveillance application categories

Category	Application
Ground-based surveillance	a ATC surveillance in airspace with radar coverage b ATC surveillance in airspace without radar coverage c Airport surface surveillance d Aircraft derived data for ground-based ATM tools
Airborne-based surveillance	a Situational awareness • Enhanced traffic situational awareness on the airport surface • Enhanced traffic situational awareness during flight operations • Enhanced visual acquisition • Enhanced successive visual approaches b Airborne spacing and separation • Enhanced sequencing and merging operations • In-trail procedure • Enhanced crossing and passing operations
Other	a Ramp control/gate management b Noise monitoring c Remote airport charges issuing d Enhanced situational awareness of obstacles e Search & Rescue (SAR), emergency response

net tools for preventing collision between aircraft or collision with terrain/obstacles are available for the controllers on the ground and for the pilot in the cockpit (Skybrary, 2011), including:

- Ground-based safety net that uses surveillance data to provide warning times of up to two minutes. Upon receiving the warning alert, controllers are expected to immediately assess the situation and take appropriate action.
- Ground-based safety net tools include:

  - Short Term Conflict Alert (STCA)
  - Area Proximity Warning (APW)
  - Minimum Safe Altitude Warning (MSAW)
  - Approach Path Monitor (APM).

- Airborne safety nets provide alerts and resolution advisories directly to the pilots. Warning times are generally shorter, up to 40 seconds. Pilots are expected to immediately take appropriate avoiding action.

- Airborne safety net tools include:

  - Airborne Collision Avoidance System (ACAS)
  - Ground Proximity Warning System (GPWS).

Radar supports ground surveillance application tools (e.g. STCA, MSAW, etc.) for the air traffic controllers to manage the airspace and aircraft separation. The emergence of new surveillance technology with higher performance in comparison to the radar system enables the possibility to implement new airborne surveillance application tools. These include the Cockpit Display of Traffic Information (CDTI), In-Trail Procedure (ITP) and Aircraft Separation Assurance System (ASAS) for improved aircraft navigation operations and self separation. In addition, new enhanced communication technologies (e.g. Mode S Extended Squitter, UAT, and VDL-Mode 4) have also enabled other airborne applications such as Traffic Information Service Broadcast (TIS-B) and Flight Information Service Broadcast (FIS-B) for enhanced situational awareness.

2.6 The current surveillance system

The current surveillance system consists of the following: Primary Surveillance Radar (PSR), Secondary Surveillance Radar (SSR), Monopulse Secondary Surveillance Radar and Multilateration. Each country or region implements the surveillance systems required by taking into account the technical performance, geographical structure, contextual environment and cost. For example, it is much more cost effective to implement Multilateration instead of SSR as both have a similar requirement on the aircraft to have a Mode S transponder on board, a mandatory requirement by ICAO. The Multilateration sensors are more flexible and cheap as they can be installed on existing infrastructures while the SSR has to be installed on a piece of land and with the associated high cost. In the case of unavailability of surveillance sources, the Air Traffic Control services are delivered using Procedural Control via radio communication.

2.6.1 Primary Surveillance Radar (PSR)

Primary Surveillance Radar (PSR) involves a beam of energy that is transmitted through an aerial and reflected back from any aircraft in its path to provide information on bearing (azimuth) and distance (range) of the aircraft (Wassan, 1994). Unfortunately, the reflections may also be from fixed objects (e.g. buildings), which tends to create clutter (Aeronautical Surveillance Panel (ASP), 2007) causing uncertainty on the target display. The PSR ground station consists of a transmitter, receiver and rotating antenna. According to ICAO, the future use of PSR in en-route airspace is expected to reduce due to its high cost and the mandatory requirement for aircraft to be equipped with a transponder to support the SSR that has the capability to supersede the PSR. However, PSR is still an important technology for security purposes in both civil and military airspace, despite its inability to uniquely identify targets, their altitude and the need for the transmission of high power pulses that limit its range. PSR remains crucial in high traffic density terminal areas to provide surveillance of aircraft not equipped with a transponder and objects on the runways or taxiways. Figure 2.2 illustrates PSR system operation.

2.6.2 Secondary Surveillance Radar (SSR)

The Secondary Surveillance Radar (SSR) sends out interrogation signals at 1030 MHz from the ground station to each aircraft within its

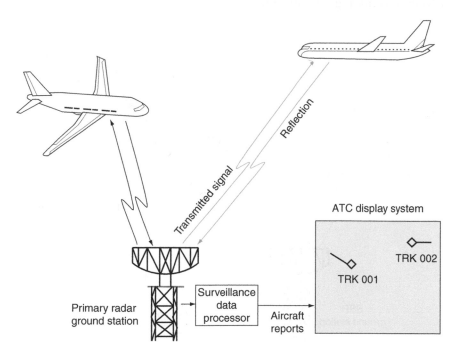

Figure 2.2 Primary Surveillance Radar.
Source: ICAO, 2007d.

range (200–250 NM) and then awaits a reply at 1090 MHz from the aircraft transponder. The aircraft's transponder responds to interrogations, enabling the aircraft's range and bearing from the ground station to be calculated independently by the SSR (ICAO, 2004b) (see Figure 2.3). The system provides an update rate of 4–12 seconds. The SSR requests two types of data from an aircraft transponder: Mode A/C or Mode S. Mode A data represents a four digit aircraft identity, while mode C data represents aircraft altitude (Wassan, 1994). Mode S is an enhancement of mode A/C by the addition of the selective addressing of targets by the use of unique 24-bit address. It also provides a two-way data link between the ground stations and the aircraft for information exchange (ICAO, 2007c).

The SSR overcomes all the drawbacks identified for PSR. However, it is unsuitable for aerodrome surface surveillance due to the accuracy limitations imposed by the transponder delay tolerance. An important achievement with SSR is the Mode S technology that will enable many new surveillance technologies to evolve in the future. For example, Multilateration technology was developed as a result of the Mode S technology evolution.

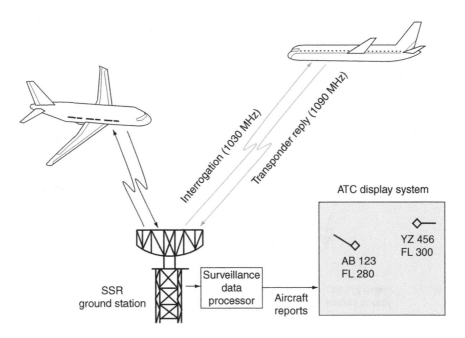

Figure 2.3 Secondary Surveillance Radar.
Source: ICAO, 2007d.

There are two classes of SSR in use currently (ICAO, 2007c):

- Classical SSR – This SSR system relies on the presence or absence of the SSR transponder replies within the beam-width. Performance can be quite poor, particularly for azimuth accuracy and resolution. This type of system is also subject to significant multipath anomalies due to the poor antenna pattern. Range accuracy depends on the variability of the fixed delay in the aircraft transponder.
- Monopulse SSR – This system measures the azimuth position of an aircraft within the horizontal antenna pattern using diffraction techniques. These techniques improve azimuth accuracy and resolution. In addition, these radars typically have large vertical aperture antennas and hence are less subject to multipath effects.

2.6.2.1 Transponder

EUROCONTROL defines a transponder as the airborne radar beacon receiver/transmitter portion of the Air Traffic Control Radar Beacon System (ATCRBS) which automatically receives radio signals from interrogators on the ground. It then selectively replies with a specific reply pulse or pulse group only to those interrogations being received on the mode to which it is set to respond (EUROCONTROL, 2011b). The transponder is mandatory equipment for the SSR operation.

2.6.2.2 Interrogation modes

The SSR has four modes of interrogation/reply: Mode A, Mode C, Mode S and intermode. Ground stations can be either Mode A/C ground stations, which can interrogate and receive replies only on Mode A/C, or Mode S ground stations, which can interrogate and receive replies on all modes. There are two classes of transponders: Mode A/C transponders and Mode S transponders. The former can respond to Mode A, Mode C and intermode interrogations only, whilst the latter can respond to all modes.

- **Mode A**
  Mode A interrogation generates a Mode A reply from the aircraft transponder which is the individual aircraft identification (also known as callsign or squawk). This identification is used by the ATC for operational purposes.

- **Mode C**

  Mode C interrogation generates a Mode C reply from the aircraft transponder, which is the encoded pressure-altitude (known as barometric altitude). Pressure-altitude is the reference for vertical separation in ICAO airspace.

- **Intermode**

  Operational compatibility between Mode S and Mode A/C aircraft and ground elements is achieved by the use of intermode (all-call communication) and by the use of the lockout protocols. Intermode transactions allow Mode S ground stations to simultaneously interrogate both Mode S and Mode A/C transponders in order to determine the Mode S aircraft. Intermode interrogations also allow the ground station to ensure that it receives replies exclusively from either Mode A/C aircraft or Mode S aircraft but not both simultaneously. The lockout protocols permit a Mode S ground station to control a Mode S transponder after its address has been determined so that it replies only to particular subsets of the possible intermode interrogations (ICAO, 2004b).

- **Mode S**

  A Mode S reply contains a 24-bit aircraft address, altitude or other data depending on the request by the ground station and aircraft capability. Mode S interrogation and replies are protected by a robust error correction scheme to ensure high reliability information is transmitted to the ground. The Mode S transponder also has the capability to report pressure-altitude in either 100 ft or 25 ft increments depending on the aircraft altimeter capability.

Mode S SSR can be categorized into two levels; Mode S "Elementary Surveillance" (ELS) and Mode S "Enhanced Surveillance" (EHS). The difference between these two levels is the amount of information given to the ATC by the SSR with the respective capabilities. Europe has issued a Mode S mandate requiring all aircraft in certain airspace to be Mode S equipped (ICAO, 2007c). The European mandate also requires Mode S ELS or Mode S EHS, to be supported.

Mode S Elementary Surveillance (ELS) provides:

- Unique 24 bit aircraft address
- SSR Mode A
- Special Position Indicator (SPI)
- Aircraft Identification (Callsign or Registration)

- Altitude with 25 ft resolution
- Flight Status (airborne or ground)
- Transponder Capability Report
- Common-Usage Ground-Initiated Comm B (GICB) Capability Report
- ACAS Resolution Advisory.

In addition to that of the ELS, the Mode S Enhanced Surveillance (EHS) provides:

- Indicated Air Speed (or Mach Number)
- Magnetic Heading
- Vertical Rate (climb, descend)
- Selected Altitude
- Ground Speed
- Roll Angle
- Track Angle Rate
- True Track Angle.

The Mode S capability provides more aircraft state information to the ATC on the ground, thereby reducing radio communication with the pilot. However, not all the information is displayed on the controller's working position. The controllers can choose the information required on the display to perform their tasks. Apart from that, the transponder technology evolution (specifically Mode S) has encouraged the emergence of new surveillance technologies such as Multilateration and ADS-B. In addition, the on-board safety net tool, ACAS, also emerged as a result of this technology.

2.6.3 Surface Movement Radar (SMR)

The Surface Movement Radar (SMR), also known as Airport Surface Detection Equipment (ASDE), is primary radar intended for aerodrome surface surveillance. Similar to the PSR technology concept, the SMR technology detects all objects within its range without uniquely identifying a target. The system provides a one-second update rate and raw digitized video to the surveillance data processor (ICAO, 2004a). The raw video is processed and displayed on an ATC screen for monitoring purposes. The system also faces signal attenuation problems during heavy rain or snow, which causes displayed targets to fade on the ATC screen.

2.6.4 Multilateration (MLAT)

The Multilateration (MLAT) technology relies on signals from an air-craft being detected by four or more MLAT ground stations to locate the aircraft. It uses the Time Difference of Arrival (TDOA) technique to establish surfaces which represent constant differences between the target and pairs of receiving stations and determines the position of the aircraft by the intersection of these surfaces (Owusu, 2003). The MLAT system is used as a surveillance tool for airport surface and terminal areas. The MLAT system requires the aircraft to be equipped with a Mode S transponder. Fortunately, this is facilitated by the mandatory requirements of ICAO for aircraft to be equipped with a transponder to support the SSR technology. The MLAT surveillance in en-route TMA is known as Wide Area Multilateration (WAM) and has coverage of 200 NM with accuracy between 10 to 500 metres. The system accuracy depends on the geometry of the target in relation to the receiving stations and also the relative time of signal receipt. The system provides an update rate of 1–5 seconds. The only disadvantage identified with this technology is the need for a minimum of four ground stations to detect the signals from an aircraft to determine its location.

2.6.5 Limitations of current surveillance systems

Based on the review of the current surveillance systems in the previous sections, the limitations of the systems are summarized:

- The PSR is unable to uniquely identify targets and their altitudes. The system requires transmission of high power pulses to detect the target, which results in low coverage.
- The classical SSR has low azimuth accuracy and resolution. It is also prone to multipath effects.
- The SMR is prone to signal attenuation which causes the target display to fade during extreme weather conditions.
- The MLAT system accuracy is dependent on the geometry of the target in relation to the receiving stations and also the relative time of signal receipt.
- All of the current surveillance systems are not suitable for remote and oceanic airspace due to difficulty in siting the sensors.

2.6.6 Procedural control

Currently, most flights are planned via intermediate way-points rather than direct routes, hence limiting the opportunity to obtain changes to cleared flight profiles. This has an adverse effect on aircraft operating costs. Flights operating outside radar and VHF coverage at present are monitored on the basis of the current flight plan (air traffic control clearance) and the pilot-reported position (air-report). The flight plan describes the assigned route along which the aircraft is expected to fly. The position reports, transmitted via HF at relatively infrequent intervals, enable the controller to monitor the aircraft's progress for conformance to its air traffic control clearance (ICAO, 1990). The application of procedural ATC ensures an adequate level of safety, at the expense of optimal flight profiles and system capacity. However, the ATC services to aircraft operating in non-radar environments employ varying degrees of automation and use different procedures for controlling traffic. As a result, pilots are required to be familiar with the individual control aspects of all flight information regions (FIRs) their flights traverse.

In order to maintain the required level of safety in the provision of ATC services, any surveillance systems deployed should co-exist with either voice or data communication service between pilot and ATC with at least similar levels of reliability (i.e. continuity) as assigned to the surveillance system. The two systems should be designed carefully so that no single point of failure can occur in both systems simultaneously. A single point of failure may occur by having a single power source for the ground remote stations or single links from the ground remote stations to the ATC centre (ICAO, 2010). In order to achieve the required level of reliability, redundant links on separate circuits need to be implemented.

Based on the review of the current surveillance technologies in this chapter, a number of advantages and disadvantages are identified and verified (based on safety data analysis and input from Subject Matter Experts (SMEs)) in Chapter 3. These disadvantages are a major obstacle to ensuring that the increasing air travel demand can be met.

2.7 Choice of surveillance technologies by Air Navigation Service Providers

The need to have surveillance systems is explained in section 2.1. However, in addition to the limitations identified in section 2.6, a number of factors and constraints impinge on the choices of surveillance technologies implemented by an Air Navigation Service Provider (ANSP) in a particular state. This can be a single surveillance technology or combination of more than one technology. The factors and constraints are outlined below (ICAO, 2007c).

2.7.1 Cost

The cost of surveillance system is a major factor in the choice of surveillance technology due to the emergence of new surveillance technologies with significantly lower cost compared to the radar system. In many states, the availability of the lower cost systems has enabled surveillance in areas where it was previously uneconomical, e.g. the implementation of an ADS-B system in non-radar or oceanic airspace whereby it is very costly to have a radar system installed and maintained. However, when the chosen surveillance technology is of a cooperative nature, the system deployment also involves airline operators and the ANSP. Therefore, the issue of who bears the cost and who benefits also needs to be considered.

2.7.2 Mixed aircraft equipage

The level of equipage for aircraft that must navigate under particular airspace constraints limits the type of surveillance technology to be deployed. Non-cooperative aircraft can only be detected by primary radar while cooperative aircraft equipped with Mode S or ADS-B capable transponders, can be detected by SSR, MLAT or ADS-B ground receivers. Consideration should also be given to general aviation and military aircraft equipage flying over the airspace.

A temporary solution for this environment is airspace segregation. The ANSP in a particular state can segregate their airspace such that equipped aircraft can operate in defined airspace while non-equipped aircraft operate in a different airspace.

2.7.3 Geography

Implementation of surveillance technologies should also take into account the obstacles to radio propagation for any particular surveillance technology in the operational area. SSR Mode S radar has a long-range capability due to its high gain antenna. It can support surveillance of upper airspace up to 250 NM if the geographical location is free from close obstacles (line of sight). Multilateration is particularly effective in areas with line of sight problems due to its ability and scalability to fill a smaller specific area of surveillance, where long-range radar is ineffective. The choice of ADS-B is not affected by geographic constraints and it fills the gaps identified in either radar or Multilateration systems to provide surveillance coverage.

2.7.4 Existing ground networking infrastructure

Availability of complete ground networking infrastructure in a particular ATC operational area makes it easier and cheaper to install ADS-B and Multilateration ground stations. This provides an advantage when considering the implementation of these two surveillance technologies. However, in the case of ADS-B, aircraft equipage also needs to be taken into account.

2.7.5 Homogeneous surveillance infrastructure

Despite the factors above, it is wise to choose new system technology from the same Original Equipment Manufacturer (OEM) of the existing ATC system, for ease of integration with the old system. This allows savings in engineering support, training, documentation management and system planning. In addition, there will be no upgrading cost of the current ATC system. For example, the Surveillance Data Processing System (SDPS) that supports multiple surveillance technologies results in both operational and cost benefits.

2.7.6 Required functionality

Different surveillance technologies may be chosen depending on the functional needs. Each technology has different functional capabilities beyond the provision of aircraft position and altitude data. Mode S radar is able to provide information of selected altitude; a Multilateration system is able to provide a precise position report independent of GPS; while ADS-B is able to provide a high update rate of a high accuracy velocity vector. In the case of ATC surveillance in dense traffic airspace, high update rate and high accuracy surveillance data are required to perform reduced separation. To support military surveillance needs, the use of primary surveillance radar is required to track unidentified aircraft in the airspace.

2.7.7 Equipage mandate

The choice of ADS-B as a surveillance system depends on a particular state's ability to mandate that for aircraft to operate in their airspace, they must be equipped with an ADS-B capable transponder or ADS-B emitter. The state's ability to issue a mandate may depend on several factors such as cost as well as on political considerations. However, the choice of SSR Mode S and Multilateration systems has no constraint of mandate, as ICAO has made it mandatory for commercial aircraft to be equipped with Mode S capable transponders (ICAO, 2007c).

2.7.8 Airspace capacity requirement

As a result of increasing air travel demand, airspace capacity needs to be increased. One method by which this can be achieved is by applying reduced separation standards and this requires high performance (accuracy, integrity, update rate) surveillance technology. At present, the separation standards are stipulated in ICAO Doc 4444 (ICAO, 2007a). However, these separation standards (3 NM and 5 NM) are based upon the utilization of PSR, MSSR, SSR Mode A/C and SSR Mode S. Hence, the choice of surveillance technology should also be based on the new surveillance technologies: ADS-B and Multilateration, which are envisioned for higher performance to support reduced separation.

2.8 Surveillance integration

When more than one surveillance system is implemented in an operational area, the surveillance data from the different sensors need to be incorporated into an ATC system for both situational awareness and any ATC separation functions. According to ICAO, these can be done in three ways (ICAO, 2007c):

- A separate ATC display for each surveillance system. However this approach is impractical for the air traffic controllers in performing their tasks.
- A priority system is displayed and other sources discarded, with the priority source providing useable data.
- A fully fused position calculation whereby data from different surveillance sensors are used to calculate the best estimate of aircraft position.

The third approach, "data fusion", is effective for airspace with redundant surveillance coverage and envisions generating higher accuracy position data. However, the integrity of each position plot from each sensor has to be carefully checked to perform the data fusion process.

Air Services Australia performed surveillance integration for airspace with MSSR and ADS-B coverage by providing different symbols on the ATC display for data from each sensor. Priority (rank) is given in this case to the sensors depending on its integrity level and availability in the surveillance area.

2.9 Separation management

Aircraft separation is performed primarily to prevent collisions and to optimize the safe use of airspace. Currently the surveillance and the communication systems play an important role in aiding air traffic controllers perform separation management on the surface and in the air. In areas without surveillance coverage, for example, in oceanic areas and remote areas, ATC is dependent on the pilots reporting their position verbally via radio frequencies. Due to uncertainty in the reported position and the low position update rate, the aircraft are separated by relatively large distances (Aeronautical Surveillance Panel (ASP), 2007). On the other hand, in the terminal area, where the surveillance systems are available, reduced separation can be performed.

However, with the increasing demand for air travel, the current surveillance systems are unable to improve the separation in order to optimize the airspace capacity. Based on analysis (Joint Planning and Development Office (JPDO)–Air Navigation Services Working Group (ANSWG), 2008) of NextGen capabilities, implementation of ADS-B across the United States National Airspace System could provide 30 per cent of capacity growth to achieve future traffic levels (three times the 2004 traffic levels). Table 2.4 shows the current separation management procedures in comparison to the future separation procedures with ADS-B in place.

2.9.1 Use of ATC surveillance system for separation

Surveillance systems such as PSR, SSR, MSSR, Multilateration and ADS-B may be used either alone or in combination in the provision of air traffic control services, including provision of separation between aircraft, provided that (ICAO, 2007a):

Table 2.4 Current and future separation management

Current separation procedures	Future separation procedures (ADS-B)
• Use single sensor reported position and correlation with primary data	• Position information from Global Navigation Satellite System (GNSS) via ADS-B
• Update rate 4.5–12 seconds	• Update rate 1–2 seconds
• Aircraft velocity is estimated from the history of reported position	• Real time aircraft velocity from modern Positioning, Navigation and Timing (PNT) system via ADS-B
	• Trajectory based operation and delegated separation procedures introduced

- reliable coverage exists in the area;
- probability of detection, accuracy and integrity of the surveillance systems are satisfactory; and
- in the case of ADS-B, the availability of the data from participating aircraft is adequate.

In addition to this, safety assessment of the surveillance systems is crucial before it can be used to provide separation. Since the focus of this book is the ADS-B system, further requirements of the system to aid aircraft separation are discussed in Chapter 4.

2.9.2　Constraints to separation minima management

In order to manage separation, a number of constraints need to be considered.

- human factors, e.g. limitation on controller workload;
- technologies (communication, navigation, surveillance), e.g. availability of the technologies;
- operational procedures, e.g. approved reduced separation minima;
- contextual environment, e.g. extreme weather conditions;
- ability of the Air Navigation Service Providers (ANSPs), e.g. to provide required resources.

The constraints caused by the limitations in the radar system are discussed further and analysed using real safety data in Chapter 3.

3 Radar system limitations

3.1 Background

Chapter 2 has reviewed the current Air Traffic Management (ATM) surveillance systems, including Primary Surveillance Radar (PSR), Secondary Surveillance Radar (SSR), Monopulse Secondary Surveillance Radar (MSSR) and Multilateration (MLAT). Surveillance systems provide Air Traffic Control Operators (ATCOs) with aircraft position in order to perform separation and airspace management, underpinned by situational awareness and functions such as trajectory prediction and conflict detection.

Currently, the primary surveillance system that supports ATM, and in particular ATC, is radar. Brooker (2004) discusses radar inaccuracies and the impacts on mid-air collision risks. His work is supported by a review of quantitative en-route safety assessment methods (Brooker, 2002) and safety assessment methodology for Communication, Navigation, Surveillance/Air Traffic Management (CNS/ATM) systems based ATC (Vismari and Camargo, 2011; ICAO, 1998c). It is clear from this assessment that radar system performance is insufficient to meet the anticipated increase in air traffic (underpinned by a reduction in separation minima) and to cater for new ATC applications (such as enhanced situational awareness for pilots and controllers, trajectory prediction and conflict detection). The radar system is thus fundamentally limited and cannot accommodate any further increase in air traffic. The limitations are particularly acute in the provision of ATC services in low altitude, remote and oceanic areas and include unavailability of service in oceanic and remote areas, limited service during extreme weather conditions and outdated equipment without spare parts (ICAO, 2000).

The impacts of these limitations manifest in the occurrence of incidents and accidents. The ICAO defines these two types of occurrences (ICAO, 2001) as below:

- An *accident* is "an occurrence associated with the operation of an aircraft, which takes place between the times that any persons board the aircraft with the intention of flight and that all such persons have disembarked, in which any person suffers death or serious injury, or in which the aircraft receives substantial damage".
- An *incident* is "an occurrence, other than an accident, associated with the operation of an aircraft which affects or could affect the safety of operation. Such incidents and accidents are reported by ANSPs to civil aviation regulators".

Automatic Dependent Surveillance Broadcast (ADS-B) (reviewed in Chapter 4) is expected to address the limitations of the radar system. Both, the FAA's Next Generation Air Transportation System (NextGen) and the European Commission's Single European Sky (SES) and its ATM Research (SESAR) programme recognize ADS-B as the key to the respective goals to modernize ATM operations and address the deficiencies in the radar systems. Unlike the radar system which operates independently, the ADS-B system relies on Global Navigation Satellite Systems (GNSS) such as Global Positioning System (GPS) to determine an aircraft's current state (location, time and related data) and derive aircraft intent information. The ADS-B system has the potential to support surveillance services not only in the terminal and en-route airspace, but also in remote and oceanic areas. Furthermore, it should enable many new surveillance applications such as synchronized situational awareness (crew-crew, crew-air traffic controller, ATC-ATC, etc.). These benefits have the potential to enhance the current operational paradigm of ATM in non-radar and remote areas without jeopardizing the required level of safety.

3.2 Methodology

Currently, there is no evidence in the public domain research literature on the analysis of the limitations of the radar system and the capability of the ADS-B system to overcome them. The literature provides the limitations in the radar system without proposing any potential solutions. Therefore, a comprehensive methodology in Figure 3.1 is developed to facilitate the resolution of the current limitations. The following sections explain the methodology in detail.

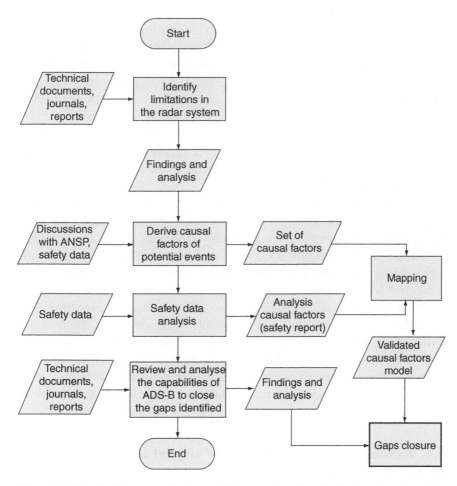

Figure 3.1 Methodology to derive, validate and resolve limitations in the radar system.

3.3 Review of the limitations

As discussed in Chapter 2, the current surveillance systems in use consist of Primary Surveillance Radar (PSR) (Wassan, 1994; Aeronautical Surveillance Panel (ASP), 2007), Secondary Surveillance Radar (SSR) (Wassan, 1994; Dawson, 2004), Monopulse Secondary Surveillance Radar (MSSR) (Dawson, 2004), Surface Movement Radar (ICAO, 2004a) and Multilateration (MLAT) (Owusu, 2003).

Each country or region implements surveillance systems taking into account technical performance, geographical structure, contextual environment and cost. For example, MLAT sensors are more flexible and relatively inexpensive as they can be installed on existing infrastructures such as high-rise buildings or communication towers while the SSR has to be sited on dedicated land which incurs space rental and higher maintenance costs. No additional cost is incurred by the airline operators as both systems have the same requirement to have a Mode S transponder on board, a mandatory requirement by ICAO (Surveillance and Conflict Resolution Systsms Panel, 2004b).

Based on the review of the current surveillance technologies in Chapter 2, a number of advantages and disadvantages are identified (Table 3.1). The disadvantages are a major obstacle to ensuring increasing air travel demand can be met.

One way to minimize the drawbacks identified in each of the current surveillance systems is to combine these technologies in a sophisticated manner (Olivier *et al.*, 2009; Bloem *et al.*, 2002) taking advantage of the strengths of each system. For example, in an area with a line-of-sight problem or limited radar coverage, integration of the radar and MLAT systems is a potential solution, taking advantage of the flexibility of installing MLAT sensors on existing infrastructure. On the other hand, in an area with multiple radar coverage, their combined usage should also enable a more reliable coverage within the radar airspace. However, the relative high cost of radar is a major barrier to the realization of this approach. In summary, although integration is useful to address some limitations, others remain, including unavailability of coverage in oceanic areas and remote regions.

3.4 Taxonomy and causal factor model for incidents/accidents

The term *system limitation* in this book refers to insufficient capability or inadequacy of a system to perform in the different phases of operation with the required performance in terms of accuracy, integrity, continuity and availability (Surveillance and Conflict Resolution Systems Panel, 2004a) which could lead to incidents or accidents. The limitations identified in the radar systems might be due to particular functional requirements overlooked during the system design phase, e.g. extreme cases such as coverage in remote areas, the need for high-performance situational awareness for flight crew and controllers, degraded visual conditions during extreme weather (Herrera *et al.*, 2009), especially for Visual Flight Rules (VFR) flights and requirements to meet future air traffic volumes. Today, no single surveillance system is available that satisfies the Required Surveillance Performance

Table 3.1 Advantages and disadvantages of the current surveillance technologies

Technology	Advantages	Disadvantages
PSR	Provides surveillance of aircraft or vehicles not equipped with transponder	• Disability to uniquely identify targets and their altitudes • Need for transmission of high power pulses that limit its range • Signal clutter • Signal attenuation • Not suitable for oceanic and remote airspace
SSR	SSR overcomes all the drawbacks identified in PSR	• False Replies Unsynchronized with Interrogator Transmissions (FRUIT) in multi-radar environments • Garbling in dense airspace • Not suitable for aerodrome surface surveillance due to the accuracy limitations imposed by the transponder delay tolerance • Dependency on transponder and on the need to have it switched on – i.e. not suitable to identify non-equipped aircraft • Not suitable for oceanic and remote airspace
SMR	Provides surveillance of vehicles not equipped with transponder and objects on the runways or taxiways	• Signal attenuation • Target not uniquely identified
MSSR	• Reduces garbling and FRUIT • Its accuracy provides for a reduction of separation minima in en-route ATC from 10 NM to 5 NM	• Dependency on transponder • Not suitable for oceanic and remote airspace
MLAT	Ability to support mode S elementary	• System accuracy depends on the geometry of the target in relation to the receiving stations and also the relative time of signal receipt • Dependency on transponder • Not suitable for oceanic airspace

(RSP) (Surveillance and Conflict Resolution Systems Panel, 2004a) required for the future traffic volumes, without jeopardizing safety.

3.4.1 Taxonomy: causal factors for incidents / accidents due to limitations in the radar systems

Definitions of "taxonomy" are inconsistent in the literature (Wilke and Majumdar, 2012). However, "scheme of classification" (Wilke and Majumdar, 2012) is adopted as the definition of taxonomy for the purpose of this book. The scheme in the context of this book refers to a set of causal factors, while the classification refers to incidents or accidents due to the limitations in the radar systems. In this book, the term "causal factor" refers to the factors that contribute to the causes of an incident or accident. The new taxonomy is derived and validated using research methods based on literature review, structured communication with Subject Matter Experts (SMEs), on the job experiences, reviews of existing taxonomies and safety report analysis. Figure 3.2 illustrates the processes applied to derive and validate the taxonomy.

In the first part of the process, an extensive literature review on the surveillance systems' functions and their limitations to support increasing air traffic was conducted and summarized (Table 3.1). The inputs used include documents on technology, system requirements and safety, from ICAO, EUROCONTROL and RTCA. The findings from the literature consist of existing required surveillance functions and those overlooked during system design and implementation to provide complete ATC services for the users in all operational contexts. Second, structured communication with a number of aviation safety personnel from ANSPs (i.e. Irish Aviation Authority, Avinor Norway and NATS UK) were surveyed to identify any limitations of the radar system (e.g. limited surveillance coverage in oceanic and remote areas) to provide complete ATC services. Third, the author's five years work experience with ATC systems operations enabled greater understanding of the radar system functions and the new functions required to accommodate future air traffic requirements. Finally, ICAO generic aviation taxonomy, Accident/Incident Data Reporting (ADREP) (ICAO, 2006a) and ESARR2 (EUROCONTROL, 2009b) safety reporting requirements were reviewed for relevance to this book. ADREP provides step-by-step guidance to generate an incident/accident report. For example, for a loss of separation incident, the following information is required to complete the ADREP reporting template:

* Minimum horizontal separation estimated (Est minimum horiz sep)
* Minimum horizontal separation prescribed (Req minimum horiz sep)

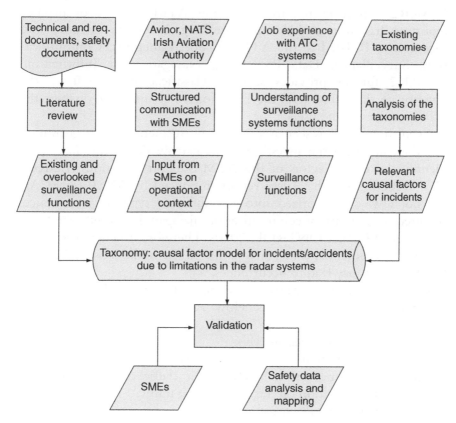

Figure 3.2 Methodology for derivation and validation of the new taxonomy.

- Minimum horizontal separation recorded (Min horiz sep recorded)
- Minimum vertical separation estimated (Est vert separation)
- Minimum vertical separation prescribed (Req vert separation)
- Minimum vertical separation recorded (Vertical sep recorded).

It is widely accepted that ADREP is a comprehensive safety reporting aid. However, it does not provide comprehensive guidance to the investigator on the analysis process. ADREP defines ATM specific occurrences as ATM or CNS service issues, including failure or degradation of all the facilities, equipment, personnel, procedures, policy and standards involved in the provision of state-approved Air Traffic Services. This classification scheme is too general for the investigator to draw any conclusion on the potential factors of a particular ATM-related occurrence. However, based on the

ESARR2 reporting requirements, the ATM specific occurrence reporting has improved through breaking down the category into Occurrence Related to the ATM Support Functions; Communication, Surveillance, Data Processing, Navigation or Information. ESARR2 further provides severity classification to indicate the severity of the ATM Support Functions that limits or disables the ATM services to the users.

Even though this is seen as a good improvement, the categorization still does not help the investigators to narrow down the analysis to identify the causal factors of the ATM Support Functions. EUROCONTROL states that safety reports produced by ANSPs in ECAC are not always categorized into the different types of occurrences required (e.g. failure of communication, navigation, surveillance functions etc.) (EUROCONTROL, 2010). Therefore, the new taxonomy provides the potential causal factors due to the limitations in the radar functions that may lead to occurrences.

Outputs from the four processes explained above feed into the derivation of the new taxonomy, as below (Ali *et al.*, 2015):

- *C1 – Lack of situational awareness*
  Factors due to the limitations in the surveillance system that caused the pilot, airline operators and/or air traffic controllers to have reduced information on the surrounding traffic.
- *C2 – Limited surveillance coverage*
  Inability to site radar in certain geographical area due to its lack of scalability resulted in some part of a particular airspace without surveillance coverage.
- *C3 – Inaccurate positioning information*
  Aircraft state vector provided by the radar to the system users differs from the actual aircraft state vector.
- *C4 – Low update rate (position data)*
  Delay (seconds) in the transmission of aircraft position by the surveillance system to the Controller Working Position (CWP). This affects the position accuracy in conjunction with the reception time at the CWP.
- *C5 – Loss of communication*
  Interruption in the air-ground and/or ground-ground communication network for data transfer.
- *C6 – Unsynchronized surveillance information between flight crew and ATC*
  Accuracy of the aircraft state vector information differs between the information displayed on board and at the CWP.

- *C7 – Visual deficiencies in extreme weather conditions*
  Inability of the surveillance system to adapt to extreme weather conditions resulting in reduced situational awareness in the particular area.

The new taxonomy only focuses on the surveillance function. Figure 3.3 shows how the new taxonomy improves the taxonomies stipulated by ADREP and ESARR2 for an ATM specific occurrence category. It is important to note that, mitigation efforts can only be proposed upon identification of specific causal factors of an event, in order to avoid repeating the safety occurrences.

In the last step of the process, the new taxonomy in Figure 3.3 and the causal model developed based on the taxonomy in Figure 3.4 are validated by surveillance and ATM safety experts from Avinor, Norway and also using safety data analysis in section 3.5. The experts' profiles are tabulated in Table 3.2.

Safety reports from ANSP are analysed to identify occurrence due to the limitations in the radar systems. The causes of the identified occurrences are further analysed to identify the contributing causal factors. The causal factors identified from the safety reports are then grouped and mapped to validate the new taxonomy (see Table 3.8).

3.4.2 Causal factor model for incidents/accidents due to limitations in the radar system

Credible incident/accident analysis requires a comprehensive taxonomy together with historical occurrences analysis information or

Table 3.2 New taxonomy validation by aviation experts from Avinor Norway

Name	Job title	Experience in ATM	Area of expertise
Bjørn Hovland	Project Manager	17 years	Surveillance systems
Trude Myhre	Senior Safety Analyst	3 years	Safety engineering
Tommy Kjelsrud	Senior Safety Analyst	7 years	Safety engineering
Kjersti Disen	Senior Safety Advisor	22 years	SMS,QMS
Anne Chavez	Safety Manager	30 years	SMS,QMS

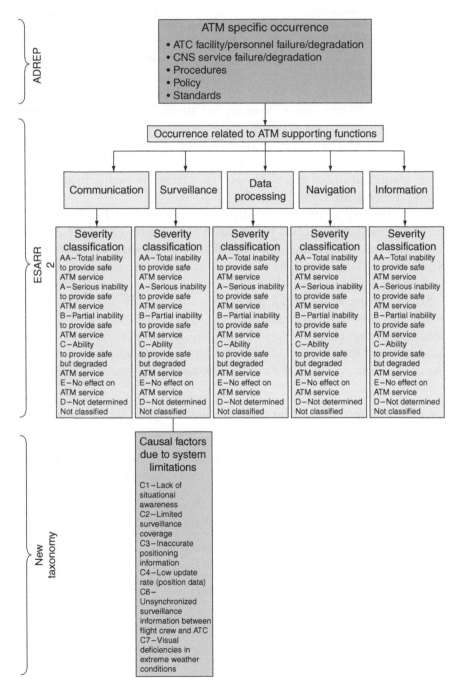

Figure 3.3 Improvement of ADREP and ESARR2 for an ATM specific occurrence category with the new taxonomy.

experience, and a causal or risk model. In order to transform the new taxonomy into an applied analytical and investigation methodology, a new causal factor model for incidents/accidents resulting from the limitations in the radar systems is developed. The new model is developed using a modified Ishikawa or fishbone diagram (Stolzer *et al.*, 2008) in Figure 3.4.

From the literature (EUROCONTROL, 2005), four main aviation incident categories – loss of separation, level bust, runway incursion and terrain incursion – are identified as the main effects of the causal factors derived. The incidents are then mapped to potential accidents; mid-air collision, ground collision and collision with obstacles. The derived causal factors are mapped to identify their potential relevance to the incidents and accidents. The method used to develop the model is reverse

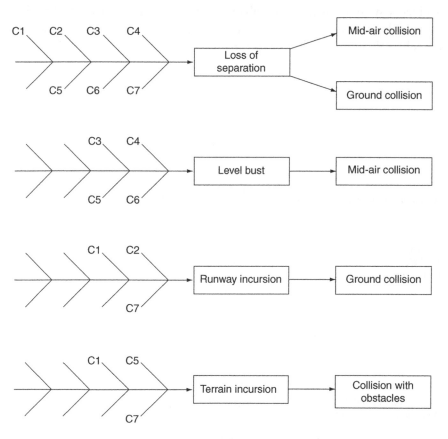

Figure 3.4 Causal factor model for incidents/accidents due to limitations in the radar systems.

engineering, whereby in the Ishikawa method, the causes of an encountered effect/problem are identified. The causal factors are derived earlier and then their potential effects are identified. The structure of the model is a fishbone. The incident and accidents are placed to the right of the spine respectively and the related causal factors are placed on the main bones extending from the spine of the diagram. Occurrence of a potential incident/accident can be due to the individual existence of any causal factor on the main bones. The model is flexible as more causal factors may be added as the subject matter evolves in the future with the emergence of new surveillance technologies.

Among others, the two main models for aviation accident-incident investigation are the Integrated Risk Picture (IRP) Risk Model (Perrin and Kirwan, 2000) and the Causal Model for Air Transport Safety (CATS) (Ale, 2009). The IRP Risk Model represents the risk of aviation accidents with a particular emphasis on the contribution of ATM. For each of the five accident categories, it embeds separate causal factors such as technical failure and human error (see Figure 3.5).

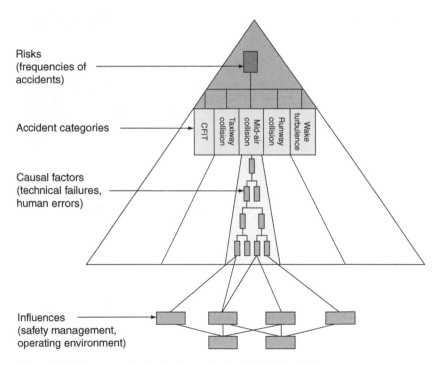

Figure 3.5 Integrated Risk Picture (IRP) Risk Model
Source: Perrin and Kirwan, 2000.

The model does not clearly mention which ATM component function may have caused the event. The risk of each accident category is quantified by providing a structured breakdown of their causes with their probability of occurrence. Although the model focuses on ATM contribution to accidents; it only considers events involving interaction with ATC.

The CATS model models causal factors (human error, technical failures, environmental and management influences) in certain characteristic accident categories including loss of control, collision and fire. The causes and consequences of such accidents differ according to the phases of flight in which they occur. In contrast to the IRP model, CATS attempts to quantify the risks of all possible aviation accidents by taking into account all possible causal factors. It considers aviation accidents to be the result of complex combinations of many different causal factors.

The IRP and CATS models are largely similar, except the scope of the CATS model is broader and the risk calculation method more detailed. The causal model proposed in Figure 3.4 represents specific causal factors for accidents/incidents due to the existing limitations in the current surveillance systems. The model differs from the IRP Risk and CATS models as these only consider technical failures of the CNS/ATM systems that may contribute to accidents, while the model proposed in this book, specifically considers the limitations of the surveillance systems that may lead to unanticipated incidents. This understanding is crucial in order to avoid unanticipated events that could jeopardize safety. Table 3.3 highlights the differences between the proposed model and the IRP and CATS models.

3.5 Safety data analysis

The aim of the safety data analysis is to validate the taxonomy identified in section 3.4 based on the analysis of comprehensive safety reports from ANSPs on accidents/incidents involving aircraft.

The key requirements for the data are:

* a detailed explanation of the causal factors of the events;
* sufficient reporting duration (to meet the analysis objective); and
* a consistent quality of reporting.

The possible outcomes of the analysis are:

* all the causal factors can be mapped on a one-to-one basis to the safety report;

Table 3.3 New causal model, IRP, CATS

Model	Application	Focus	Output	Focus on ATM supporting function limitations
IRP	Accidents with ATM contribution that involves ATC interaction. In particular mid-air collision, runway collision, taxiway collision, CFIT and wake turbulence accident	Each accident can be explained by either single or combination of causal factors	Qualitative and quantitative	None
CATS	All accidents	Complex combination of all possible causal factors	Qualitative and quantitative	None
New causal factor model	Incidents and accidents due to limitations in the surveillance system function	Each incident can be explained by either single or combination of causal factors	Qualitative	Focus on surveillance function only

- only part of the causal factors can be mapped to the safety reports; or
- the causal factors are insufficient to cover all the factors identified in the safety reports.

The analysis involves a step-by-step approach to identify the causal factors of occurrences due to the limitations in the radar system. The steps are described as follows (Ali *et al.*, 2015):

Step 1 – Reports of occurrences involving aircraft in the radar airspace are extracted from the database.

Step 2 – The occurrences due to ATM supporting functions, which include navigation, surveillance and communication systems are reported as "Insufficient separation", "Lack of or reduced ability to provide ATM Services" and "Inability to provide Air Traffic Services (ATS)". The occurrences under these categories are identified.

Step 3 – Narratives of all the reports identified in Step 2 are analysed to identify the specific ATM supporting function that contributed to the particular occurrence.

Step 4 – Causal factors of occurrences associated with the surveillance function are placed into three categories, based on the narratives: "Contextual Environment", "Human Error" and "System Limitation".

Step 5 – Percentage of occurrences for the three categories in Step 4 are measured.

Step 6 – Questionnaire is developed and given to the ANSP experts to identify the underlying causes of all the occurrences under the "System Limitation" category. The questionnaire was developed based on the statistical analysis results.

Step 7 – The narratives of the occurrences under the "System Limitation" category are analysed further to identify the specific causal factors of the occurrences.

Step 8 – Causal factor's grouping phrases are developed to group similar causal factors reported using different sentences in the safety reports by different investigators.

Step 9 – Number of occurrences for each causal factor identified in the safety reports are measured and mapped to the corresponding grouping phrase developed in Step 8.

After reviewing safety reports from various ANSPs and regulators, it was found that data from the Norwegian ANSP, Avinor, was best suited for this analysis due to its completeness and organized structure. In addition, structured communication with SMEs from EUROCONTROL indicated that, based on their working experience with the European countries on safety issues, the Norwegian ANSP, has an excellent reputation for reporting safety occurrences. All the safety reports are stored in the MESYS database, which contains original reports and the findings of investigations. This reporting system complies with the EUROCONTROL Safety and Regulatory Requirements (ESARR2) (EUROCONTROL, 2009b). Based on these facts, the organization was evaluated as a reliable source of reporting and five years (2008–2012) of incident data were gathered accordingly.

The Norwegian ANSP operates 46 airports in Norway, 12 of these in cooperation with the armed forces. Their operations also include air traffic control towers, control centres and technical infrastructure for aircraft navigation and surveillance. Figure 3.6 shows the radar locations maintained by the ANSP. Most of Norway's airspace has redundant radar coverage. Table 3.4 presents Instrument Flight Rules (IFR) and Visual Flight Rules (VFR) aircraft movements for all the 46 airports in Norway for the period 2008–2012. The trend indicates a gradual increase from 2009 to 2012 after a significant drop from 2008 to 2009.

Figure 3.6 Radar locations for Norwegian airspace.
Source: Avinor, 2011.

Table 3.4 IFR and VFR traffic for Norwegian airspace (2008–2012)

Year	Traffic
2008	870 365
2009	834 883
2010	841 859
2011	869 348
2012	893 813

3.5.1 Safety data analysis results

A descriptive statistical analysis (Table 3.5) on the five years of safety data shows that with the exception of 2010 and 2011, with the same level of incidents, the number of incidents has been increasing significantly. However, the number of accidents decreased significantly from 2010 to 2011 (from nine accidents to three) despite the increase in air traffic. The accident figure increased again from 2011 to 2012 by four accidents.

From the analysis of safety reports, three main causes of incidents/accidents (known as occurrence type in the safety report and severity classification in the ESARR2 taxonomy) can be directly associated to occurrences related to the ATM functions (defined in Figure 3.3):

- *Insufficient separation*
  In the absence of prescribed separation minima, a situation in which aircraft were perceived to pass too close to each other for pilots to ensure safe separation (EUROCONTROL, 2009b).
- *Lack of or reduced ability to provide ATM services*
  An event in which elements in the ground ATM system performances are unserviceable, insufficient, unavailable or corrupted so that

Table 3.5 Incident, serious incident and accident for the year 2008–2012 (Norwegian airspace)

Occurrence class	2008	2009	2010	2011	2012
Accident	3	4	9	3	7
Incident	882	1391	1506	1505	1775
Serious incident	10	9	9	10	13
Total	895	1404	1524	1518	1795

Source: Ali *et al.*, 2015

the safety of traffic, ensured through the provision of air navigation services, is impaired or prevented (EUROCONTROL, 2009b).

• *Inability to provide Air Traffic Services (ATS)*
  An event in which elements in the ground ATS system are unavailable (EUROCONTROL, 2009b).

These categories are defined in detail in ESARR2 (EUROCONTROL, 2009b) and the MESYS database design is in line with the ESARR2 template. Figure 3.7 presents the occurrence type and number for the period 2008–2012.

Occurrences caused by "insufficient separation" were at the lowest level in 2009 (5 incidents) and highest in 2012 (25 incidents). The incidents caused by "lack of or reduced ability to provide ATM services" were at the highest level in 2009 (17 incidents) and lowest in 2008 and 2011 (1 incident each year). Overall, the "insufficient separation" category has the highest count over the five years (77 incidents). There is no significant pattern shown in the occurrence of incidents due to "lack of or reduced ability to provide ATM services". However, incidents caused by "inability to provide Air Traffic Services" and "insufficient separation" show a continuous increase from 2008 to 2010 and 2010 to 2012 respectively. No incidents were

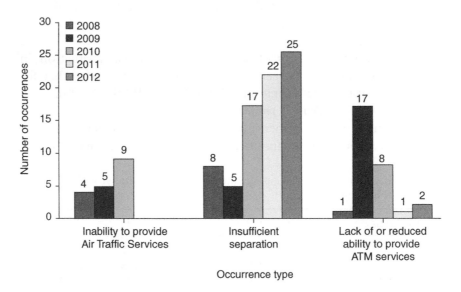

Figure 3.7 Occurrence number and type for 2008–2012.
Source: Ali *et al.*, 2015.

recorded in 2011 and 2012 for the "inability to provide Air Traffic Services" category. The statistics in Figure 3.7 include occurrences related to all the ATM supporting functions. However, the occurrences of interest in this book are only those related to the surveillance function. A review of the safety reports is used to refine the analysis and identify the occurrences related to surveillance function.

Figure 3.8 shows the number of occurrences related to surveillance functions and the other ATM supporting functions. The figure shows that of the 18 occurrences from "inability to provide Air Traffic Services", 10 are due to the surveillance function. Of the 77 occurrences from "insufficient separation", 55 are due to the surveillance function and of the 29 from "lack of or reduced ability to provide ATM services", 11 are due to the surveillance function. The other ATM supporting functions are communication, data processing, navigation and information functions as stated by ESARR2.

Based on the narratives for the occurrences associated with the surveillance function identified in Figure 3.8, the causal factors for each occurrence type are identified and categorized as follows:

- *Contextual environment* – External air transport environment includes the physical environment outside the immediate work area such as weather (visibility/turbulence), terrain, congested airspace and

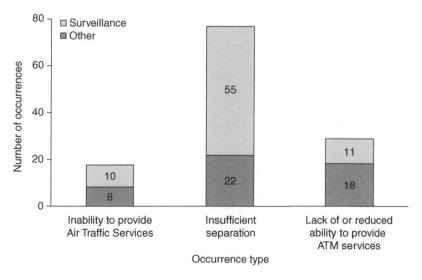

Figure 3.8 Occurrence related to ATM supporting functions.
Source: Ali *et al.*, 2015.

physical facilities and infrastructure including airports as well as broad organizational, economic, regulatory, political and social factors (ICAO, 1993). This definition is adopted in this book.

- *Human error* – Literature search and structured communication with a safety expert from EUROCONTROL, Peter Stastny, and the ATM Safety Researcher Lucio Vismari from Sao Paulo University revealed that ICAO does not have an agreed definition for Human Error. Therefore, the nearest definition relevant to this research is by Isaac and Ruitenberg (1999) in which human error in ATM/ATC is defined as "intended actions which are not correctly executed". Hollnagel *et al.* (1995) refined this definition by adding that, "human error, can denote a cause, as well as an action". This refined definition is adopted in this book.
- *System limitation* – Defined in section 3.4.

Table 3.6 shows the percentage of causal factor categories for each occurrence type related to the surveillance function, over the five years. The results show that 80 per cent of "inability to provide Air Traffic Services" occurrences are due to "system limitation". For "insufficient separation", 45.45 per cent of the causal factors are categorized as "system limitation" with "human error" being 52.73 per cent. The corresponding figures for "lack of or reduced ability to provide the ATM services" occurrences are 54.55 per cent and 45.45 per cent. In summary, the results show that "system limitation" is the highest contributing factor to the occurrences due to "lack of or reduced ability to provide ATM services" and "inability to provide Air Traffic Services".

A set of questions were raised to verify and identify the underlying causes of the findings in Figures 3.7 and 3.8 and Table 3.6 from an

Table 3.6 Categorization of causal factors for occurrence related to surveillance function

Occurrence type	Categorization of causal factors		
	Contextual environment (%)	Human error (%)	System limitation (%)
Lack of or reduced ability to provide ATM services	0	45.45	54.55
Inability to provide Air Traffic Services	10.00	10.00	80.00
Insufficient separation	1.82	52.73	45.45

Table 3.7 Questionnaire to Avinor

Questions	Response from Avinor
The analysis in Figure 3.7 shows continuous increase in occurrence due to "Inability to provide ATS" (2008–2010). Further analysis in Table 3.6 shows that 80% of the causes for the occurrences are due to "System limitation" related to the surveillance function. What are the plausible reasons for this? However, this significantly improved to zero occurrences in the last two consecutive years, 2011 and 2012. What are the plausible reasons for this drastic improvement?	Before 2010, technical issues involving ATC systems were considered as occurrences, and hence logged in the MESYS database. However, after 2010, they were no longer recorded in the MESYS. This may have been the justification for the improvements seen in 2011 and 2012. In addition, there was no radar fall out experienced in the two years. Avinor feels it has more stable technical systems in the past few years. A number of occurrences of radar fade out were experienced within the years 2008–2010, particularly in the north of Norway.
Figure 3.7 shows a continuous increase from 2008 to 2012 except for a small decrease in 2009 for occurrence due to "Insufficient separation". Based on the analysis in Figure 3.8, 70% of the occurrences are related to the surveillance function. Further analysis on the 70% of the occurrences, indicated that 52.7% are due to "Human error" and 45.45% are due to "System limitation". What is your view on this?	No comments were given on the human errors found; instead, Avinor stated that the figures found should reflect the error. Avinor added that any degradation of a system that supports separation services to aircraft results in larger aircraft separation. Hence, this is also considered as an occurrence.
The number of occurrences due to "Lack of or reduced ability to provide ATM services" has been decreasing from 2009 to 2012. What is the plausible reason for this improvement?	Avinor stated that the following reasons may have contributed to the improvements seen: • more stable systems in the past few years; • exclusion of technical issues from MESYS; • under-reporting in some units; and • no automatic comparison of reporting with "actual" incidents.

Source: Ali *et al.*, 2015.

Table 3.8 Mapping of causal factors (safety report) to derive causal factors

Causal factors identified from safety reports (Norwegian ANSP 2008–2010) for "occurrences" due to "system limitation"	Number of occurrences	Derived causal factors as the consequences of limitations in the current surveillance systems
1 Short Term Conflict Alert (STCA) not triggered due to high speed of aircraft	1	C1 – Lack of situational awareness
2 Lack of ATC awareness on traffic	3	
3 Lack of pilot awareness on surrounding traffic	3	
1 No situational awareness in certain oceanic sectors due to unavailability of radar coverage	3	C2 – Limited surveillance coverage
2 VFR aircraft not detected	2	
3 No radar for operational use in certain areas due to difficulty to site radar	3	
4 Aircraft not visible on a specific radar display due to a line-of-sight problem between aircraft antenna position and radar station when aircraft is in high altitude	2	
5 No ATC surveillance coverage in uncontrolled airspace	1	
6 No radio communication and track identification on radar display	2	
1 Inaccurate track display on radar screen	1	C3 – Inaccurate positioning information
2 Inaccurate information on aircraft speed	1	
1 Delay on track update on radar display	3	C4 – Low update rate (position data)
1 Radio communication problem	2	C5 – Loss of communication
2 Unavailability of radio coverage in certain oceanic and remote areas	2	
3 Radio frequency congestion	1	
1 Unsynchronized traffic information between TCAS (aircraft) and ATC	3	C6 – Unsynchronized surveillance information between flight crew and ATC
2 Misinterpretation of actual aircraft position – ATC	2	
3 Unsynchronized situational awareness between pilot and ATC	2	
1 Surface Movement Radar (SMR) provides poor visibility of track during extreme weather condition	1	C7 – Visual deficiencies in extreme weather conditions
2 Unavailability of limited radar coverage due to extreme weather condition	1	

Table 3.9 Example scenario (Avinor safety report) that shows correlation between system limitation and human error

Ocurrence	Ocurrence type	Causal factor identified from safety report	Causal factor categorization	Analysis based on the report narrative
Incident	Insufficient separation	Unsynchronized traffic information between TCAS (aircraft) and ATC	System limitation	Due to the effect of limitation in both TCAS and ATC surveillance systems, there is a tendency for the pilot or the controller to make an error. The ATC might give a wrong separation instruction based on the radar estimation or the pilot may navigate at wrong separation due to TCAS alert

Table 3.10 Analysis of the Ueberlingen accident

Occurrence	Occurrence type	Causal factor identified from safety report	Causal factor categorization	Analysis based on the report narrative
Accident	Insufficient separation	Unsynchronized traffic information between TCAS (aircraft) and ATC instructions	System limitation	The imminent separation infringement was not noticed by ATC in time. The instruction for the TU154M to descend was given at a time when the prescribed separation to the B757-200 could not be ensured anymore. The TU154M crew followed the ATC instruction to descend and continued to do so even after TCAS advised them to climb. This manoeuvre was performed contrary to the generated TCAS RA.

operational point of view. Structured communication with five ATM personnel (Table 3.2) was conducted on 24 March 2013 at Avinor, Norway to discuss the questions. The results are presented in Table 3.7.

The safety data analysis and the questionnaire indicate that the incidents identified are mainly related to the limitations in the radar system and that the incident numbers are also partially influenced by the reporting culture.

Therefore the incidents identified under the "system limitation" category (Table 3.6) from the safety reports, are analysed further to extract their causal factors, and then grouped and mapped to the causal factors identified in section 3.4 in Table 3.8.

Another important finding from this analysis is that some causal factors identified from the safety reports categorized as due to "system limitation" have a correlation with "human error". In such cases, the effect of system limitation may lead to human error. Tables 3.9 and 3.10 show examples of this scenario. It is not possible to show the correlation statistically as the variables of correlation are based on the interpretation of the safety reports.

This scenario can also be associated with the Ueberlingen accident.

> On 1 July 2002 at 21:35:32 hrs a collision between a Tupolev TU154M, which was on a flight from Moscow/Russia to Barcelona/Spain, and a Boeing B757–200, on a flight from Bergamo/Italy to Brussels/Belgium, occurred north of the city of Ueberlingen (Lake of Constance). Both aircraft flew according to IFR (Instrument Flight Rules) and were under control of ACC Zurich. After the collision both aircraft crashed into an area north of Ueberlingen. There were a total of 71 people on board of the two airplanes, none of which survived the crash.
> (German Federal Bureau of Aircraft Accidents Investigation, 2004)

The analysis of the Ueberlingen accident in Table 3.10 shows that the unsynchronized surveillance information from ATC with the TCAS system led to the human error which finally caused the catastrophic event. The incident would have been avoided if both the pilot and ATC had the same level of situational awareness of the traffic.

In summary, the analysis in this chapter shows that the limitations in the radar systems are significant as these constitute the main contribution to all the occurrences related to the surveillance function. Hence, improved and high-performance surveillance systems are required to support increasing air traffic based on the performance required by ICAO. The next chapter describes the ADS-B system in detail.

4 Automatic Dependent Surveillance Broadcast (ADS-B)

4.1 Background

As discussed in Chapter 1, ICAO endorsed the Future Air Navigation System (FANS) CNS/ATM concept, which is largely based on the satellite technologies in 1992. ADS-B is one of the enablers of this concept, aimed to improve airspace capacity. It is envisioned to overcome the limitations of the radar system (discussed and verified in Chapter 3) and to modernize the ATM system. It is the key driver of the Single European Sky (SES) ATM Research (SESAR) and the USA Next Generation Air Transportation System (NextGen) programmes.

4.2 Principle of ADS-B operation

RTCA (2002) defines ADS-B as a function on an aircraft or a surface vehicle operating within the surface movement area that periodically broadcasts its position and other information without knowing the recipients and without expecting acknowledgements as the system only supports one-way transmission. The system is automatic in the sense that it does not require external intervention to transmit the information. It is characterized as dependent due to its dependence on aircraft navigation avionics to obtain the surveillance information. ADS-B is a cooperative system, because it requires common equipage for aircraft, or vehicles on the airport surface to exchange information. It provides aircraft state information such as horizontal position, altitude, vector, velocity and trajectory intent information. The last of these is critical for trajectory prediction which is the basis of the trajectory-based operations concept of SESAR and NextGen.

The ADS-B system architecture is divided into two subsystems, "ADS-B Out" and "ADS-B In". ICAO (2003b) defines the term

"ADS-B Out" as the broadcast of ADS-B transmissions from the aircraft, without the installation of complementary receiving equipment to process and display ADS-B data on the cockpit displays. The complementary subsystem is "ADS-B In", which provides air-to-air situational awareness to the pilots. ADS-B Out has the capability to operate independently to provide air–ground surveillance services to the ATC. On the other hand, implementation of ADS-B In requires fully operational ADS-B Out as a pre-requisite, certification of cockpit displays, consideration of pilot human factors and other activities that will have a longer deployment schedule.

An ADS-B equipped aircraft uses an on-board navigation system to obtain the aircraft position from GNSS. The system then broadcasts periodically the position, velocity and intent data to other ADS-B equipped aircraft and ADS-B ground stations within its range via a data link service. The ground stations transmit the received ADS-B reports to a surveillance data processing system to process the data for ATC use. Figure 4.1 illustrates the ADS-B system.

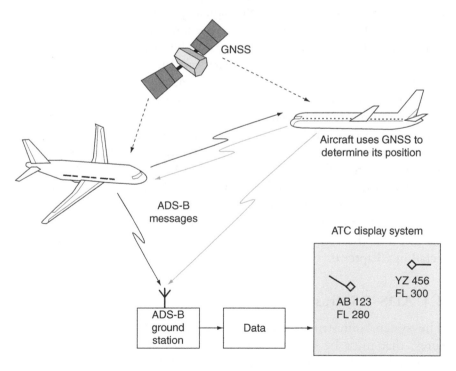

Figure 4.1 Automatic Dependent Surveillance Broadcast (ADS-B).
Source: Ali, 2016.

4.3 Key differences between radar and ADS-B

The main difference between radar and ADS-B surveillance is the means of determining the aircraft position and state vector information. As discussed in Chapter 2, radar determines the position from the travel time of a ground-emitted beam reflected by the aircraft and detected by the ground-based station, independent of any aircraft systems. Aircraft speed, direction, turn rate and other state vector information are estimated from consecutive position reports. ADS-B uses position information and state-vectors computed on board by the aircraft navigation system and broadcast this information via data link. The altitude information is obtained in both cases from an air data computer or barometric altimeter on board the aircraft. The advantages and disadvantages (Aeronautical Surveillance Panel (ASP), 2007; ICAO, 2007c; Comsoft GmbH, 2007) of the ADS-B system are summarized in Table 4.1.

The disadvantages identified in Table 4.1 may contribute to failures of the ADS-B system. The failures may lead to either corruption or loss of ADS-B data. For example, signal spoofing may cause the on-board navigation receiver to estimate the aircraft position to be somewhere other than where it actually is or to be located where it is but at a different time. The corrupted navigation data will be transmitted to the ADS-B transponder, which will subsequently be broadcast to the users. This may impose a safety threat to the ATC and other aircraft navigation operations which rely on the ADS-B system. In addition, failure of the on-board navigation system may lead to loss of ADS-B position information to the users. Ali *et al.* have done rigorous work to investigate potential ADS-B failure modes (Ali *et al.*, 2014). Apart from that, a security issue is also foreseen due to the nature of the ADS-B system, which broadcasts aircraft information to all equipped recipients. This allows tapping of the surveillance information with harmful intention to the aircraft. The potential security issues are discussed in detail in Chapter 6.

4.4 ADS-B infrastructures

The system infrastructure includes ground and airborne infrastructures, that must be certified based on the ICAO/RTCA standards. Figures 4.2 and 4.3 illustrate the avionics for ADS-B Out and ADS-B In. Tables 4.2 and 4.3 show the ground and airborne components, their functionalities and the required standards. The aircraft equipage

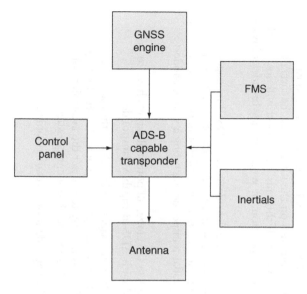

Figure 4.2 ADS–B Out avionics.
Source: Ali, 2016.

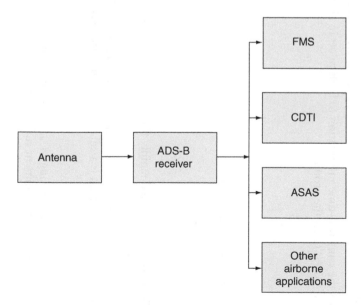

Figure 4.3 ADS–B In avionics.
Source: Ali, 2016.

Table 4.1 Advantages and disadvantages of ADS–B system

Advantages	Disadvantages
Provides the same real-time information to both pilots in aircraft cockpits and ATC on the ground.	As with any secondary surveillance technology, successful surveillance requires the cooperation of the targets. However, ADS–B not only relies on a functional transponder, but also on the integrity of the aircraft navigation system. If this fails, the aircraft will not be able to broadcast its position, or worse, it may broadcast invalid positions.
Enables efficient airspace usage.	It is relatively easy to broadcast fake ADS–B messages simulating non-existent aircraft. Both of these cases are broader, but not substantially different in risk from a classical secondary radar transponder reporting a wrong Mode–C altitude.
ADS–B can be implemented rapidly for a relatively low cost compared to radar.	For complete coverage, all potential targets have to be equipped with ADS–B capable transponders.
Provides a much greater margin in which to implement conflict detection and resolution than is available with any other system.	Since ADS–B messages are broadcast, they are available to all aircraft with the right equipment. Except for regulatory action, there is no way to restrict the availability of aircraft positions. This may lead to security issues.
Enhanced aviation safety through features such as automatic traffic call-outs or warnings of imminent runway incursion.	It is totally dependent on aircraft avionics.

ADS-B technology can be scaled and adapted for use in general aviation, ground vehicles and in airspace where radar is ineffective or unavailable.	The service outages are expected due to poor GPS geometry when satellites out of service.
General aviation (GA) aircraft can use ADS-B datalinks to receive flight information services such as graphical weather depiction and textual flight advisories. In the past, these services have been unavailable or too expensive for widespread use in GA.	The ground receiver sensors require optimum site with unobstructed view to aircraft.
Reduced cost for ground station maintenance. According to the FAA (FirebirdV8, 2006) the maintenance cost for a radar station costs as much to maintain 20 Ground Based Transceivers (GBT) for ADS-B.	Signal jamming due to the use of same frequency (Mode S) by many systems such as SSR, TCAS, MLAT and ADS-B, particularly in dense airspace.
Enables pilot to receive traffic information with aircraft identification, direction and relative altitude.	
Provides ability to change observation area along with option to take closer look at individual aircrafts.	
Reduces ATC workload.	
Increases situational awareness of the pilot.	

Table 4.2 ADS-B ground infrastructure components

Components	Function	Standard
ADS-B receiver and antenna	Receive ADS-B messages broadcast from aircraft	ED-129 Technical Specification for 1090 MHz Extended Squitter ADS-B Ground Station
GPS clock	Time stamp received ADS-B messages	ICAO Annex 10
Communication link	Transmit message from ground station to ATC surveillance data processing unit	Any form of secured dedicated private network connection such as lease line, fibre optic.
ADS-B situational display	Display aircraft position and state vector in a similar manner as radar	Controller Working Position (CWP) standard similar to radar displays

requirement for ADS-B differs according to the implementation of ADS-B Out and ADS-B In. The operational ADS-B Out is a pre-requisite for the ADS-B In implementation. Table 4.4 presents the requirements for both ADS-B Out and ADS-B In.

To enable an ADS-B equipped aircraft to operate in non-radar airspace, it has to be certified to Acceptable Means of Compliance (AMC) 20–24 standard (EASA, 2008) (airworthiness and operational approval of the "Enhanced Air Traffic Services in Non-Radar Areas using ADS-B Surveillance"). According to (Rekkas, 2013) the EU Regulation 1207/2011 applies to enable aircraft to operate in radar airspace. However, this regulation is not adopted globally.

According to the Surveillance Performance and Interoperability Implementing Rule (SPI-IR) (EUROCONTROL, 2011a), aircraft with airworthiness certification on or after 8 January 2015 are to be equipped with secondary surveillance radar transponders with the capabilities specified in Annex IV of the SPI-IR (*forward-fit*), while aircraft with airworthiness certification before 8 January 2015 are to be equipped with the same by 7 December 2017 (*retrofit*). The requirements stipulated in Annex IV of the SPI-IR enable full implementation of ADS-B Out by the aircraft operators. However, to date no regulation has been placed on the Air Navigation Service Providers (ANSPs) to implement ADS-B ground infrastructures. It is expected that this rule will follow in the near future due to the need to implement ADS-B globally.

Table 4.3 ADS–B airborne infrastructure components

Component	Function	Standard
Navigation source (e.g. GPS)	Derive and transmit aircraft position and state vector information to ADS–B emitter	• TSO C-129A, TSO C-129 or TSO C-129A; or • TSO C-145/C-146 or TSO C-145A/C-146A
ADS–B avionics (standalone box for UAT) *For 1090ES ADS–B, this function will be performed by the transponder (MODE S ES transponder)	Encode and broadcast ADS–B message	TSO–154c (UAT) TSO–166b (1090ES)
Antenna for ADS–B, transponder, GPS	To support ADS–B OUT, a single antenna at the bottom of the aircraft is required. To support ADS–B IN, antenna diversity is required, whereby two antennas – one at the top and one at the bottom of the aircraft – are required	TSO–154c (UAT) TSO–166b (1090ES)
Antenna duplexer	To enable one antenna sharing between transponder and ADS–B unit	
Barometric altimeter	Generate aircraft altitude	
Altitude encoding altimeter	Synchronize altitude information transmitted in the ADS–B message and altitude transmitted by the transponder	
Control panel	To enable pilot to key in or select aircraft identification, SPI, emergency pulse	
Flight Management System (FMS)	To manage flight plan and connect to other avionics	

Table 4.4 Aircraft equipage requirements for ADS–B Out and ADS–B In

ADS-B Out	ADS-B In
Precision GPS source – standards vary worldwide but enhanced standards such as TSO–C145 or TSO–C146 will work anywhere	ADS–B Out system
A transmitting radio – e.g. ADS–B qualified Mode S transponder (1090ES) or dedicated ADS–B data radio (978 MHz), only used below 18K feet in US airspace	TIS–B data transmission at both 1090ES and 978 MHz
Simple controls – module to enter a squawk code and verify that ADS–B is working	FIS–B graphical weather data is transmitted only at 978 MHz. The system is smart enough to know the exact location and provides a total weather map with prioritization around the current position
Data link – 1090ES data link is commonly referred to as "Mode S Extended Squitter" or 978 MHz data link is commonly referred to as "UAT (Universal Access Transceiver)"	Receiver or transceiver for ADS–B In function

It is noted that ADS-B is highly dependent on navigation and communication systems. At present GPS is widely used and supported as the navigation source for ADS-B. Three types of data links are used in states; Mode S Extended Squitter (1090ES), Universal Access Transceiver (UAT) and VDL-Mode 4 of which 1090ES is the most widely deployed.

4.4.1 On-board navigation source for ADS-B

This section will focus on the GPS, at present the most widely used navigation source for ADS-B. However, it should be noted that there is no mandate on using GPS as the navigation source. Any on-board navigation system which satisfies the required standards is acceptable.

4.4.1.1 Requirements

The minimum requirements for the navigation source for ADS-B are specified in the Technical Standard Orders TSO-C129, TSO-C145, TSO-C146 and TSO-C196. According to the RTCA, all GPS equipment compliant with the "Minimum Operational Performance Standards for Global Positioning System/Wide Area Augmentation System Airborne Equipment" are expected to satisfy the requirements for ADS-B applications.

The navigation source should have the capabilities to provide position accuracy (e.g. Horizontal Figure of Merit – HFOM) and integrity (e.g. Horizontal Protection Level – HPL) (Federal Aviation Administration, 2010a). In addition, the integrity level should be equal to or less than 10^{-7} per flight-hour with integrity time to alert equal to or less than 10 seconds. The system is also required to be at least compatible with GNSS receivers that perform receiver autonomous integrity monitoring (RAIM) and fault detection and exclusion (FDE), along with the output of corresponding measurement status information, as well as integrity containment bound and 95 per cent accuracy bound indications (EUROCONTROL, 2011a).

4.4.1.2 Architecture

GPS can be divided into three elements: ground, space and user segments (Royal Academy of Engineering, 2011):

- The ground or control segment is used to upload data to the satellites, to synchronize time across the constellation and to track the satellites to enable orbit and clock information.

- The space segment consists of the GPS satellites in six orbital planes. Twenty-four satellites make a full constellation, although as of November 2013 there are 31 in active service (US Air Force, 2013).
- The user segment consists of the receivers and associated antennae, to receive and decode the signal and compute Position, Navigation and Time (PNT) and related information.

4.4.1.3 Measurements

GPS is a ranging system with three carrier frequencies, all multiples of a fundamental frequency (Table 4.5). The distance is derived primarily by measuring the time difference between the transmission of a coded signal from the satellite and reception at the receiver. This range is known as the pseudorange rather than range since it includes a number of system unknowns such as clock biases and propagation delays which must be solved for or estimated. The carrier phase of the signals can also be used to derive the range, providing a more accurate position, but with inherent ambiguity. Ranges to at least four satellites are required to determine position and time.

The navigation message is transmitted from the satellite to the user and gives the satellite identifier together with information on satellite health, predicted range accuracy, ionosphere and clock correction coefficients as well as orbital ephemeris to allow the receiver to calculate the satellite position. The message also contains an almanac which gives status, location and identifier information for all satellites in the constellation.

4.4.1.4 Backup

In the case of unavailability of GPS, Inertial Navigation Systems (INS) can be used as a backup positioning source for ADS-B. INS, also known as Inertial Reference Unit (IRU), is an independent system

Table 4.5 GNSS RF carrier frequencies

GPS frequencies
L1 1575.42 MHz
L2 1227.60 MHz
L5 1176.45 MHz

Source: Royal Academy of Engineering, 2011.

comprising gyros and accelerometers that provide aircraft state, position and velocity information in response to signals resulting from inertial effects on the system components. Once initialized with a known position and heading, INS continuously calculates the aircraft position and velocity. However, there is currently no requirement to provide a backup navigation source for the ADS-B system.

4.4.2 Data link technologies for ADS-B

Data link technologies enable uplink and downlink of data between aircraft and ground-based ATC. Three types of potential ADS-B data links are proposed (ICAO, 2003c): the Mode S Extended Squitter (1090ES), the Universal Access Transceiver (UAT) and the VHF Digital Link (VDL) Mode 4. The data link characteristics are discussed in the following sub-sections and their differences are summarized in Table 4.6. Mode S 1090ES is explained in greater detail in Section 4.5, given its mandate by ICAO as the global data link for ADS-B.

4.4.2.1 Universal Access Transceiver (UAT)

The Universal Access Transceiver is a bi-directional data link developed to support ADS-B. It also supports the Flight Information

Table 4.6 Comparison between 1090ES, VDL4 and UAT

1090ES	VDL4	UAT
Single channel	Multi channel	Single channel
Frequency 1090 MHz	Frequency 108–137 MHz	Frequency 978 MHz
Random access	Time slot access	Time slot access
Fixed and limited channel data bandwidth	Bandwidth 19.2 kbps	Bandwidth 2–3 MHz
Fixed ADS-B reporting rate	Variable ADS-B reporting rate	Fixed ADS-B reporting rate
Extension to Mode S	New system	New system
Support air–air broadcast, uplink, downlink	Support air–air broadcast, uplink, downlink	Support air–air broadcast, uplink, downlink
ICAO standard exists	ICAO standard exists	ICAO standard exists
Mandatory equipment	Test equipment	Test equipment

Service Broadcast (FIS-B) such as weather and flight service information for aircraft. In addition UAT is capable of supporting transmission of radar information via the Traffic Information Service Broadcast (TIS-B) to ADS-B equipped aircraft. This enables the provision of situational awareness of unequipped aircraft. The data link utilizes the 978 MHz frequency. UAT data link networks are being installed as part of the FAA's NextGen, typically for general aviation users.

4.4.2.2 VHF Digital Link (VDL) Mode 4

VHF Digital Link Mode 4 (VDL4) is a digital data link designed to operate in the VHF frequency band using one or more standard 25 KHz VHF communications channels. It is capable of providing digital communications and surveillance services between aircraft and ground vehicles, as well as ground stations (EUROCONTROL, 2001–2013). VDL4 is based on the Self Organizing Time Division Multiple Access (STDMA) technology. This concept allows VDL4 to operate without a centralized co-coordinating station, thus eliminating the need for ground infrastructure. However, ground stations serve an important role in providing other services that enhance VDL4 operations. VDL4 supports broadcast and point-to-point communication with a minimum of overhead information, essential for time critical data exchange and low-end users.

4.4.2.3 Mode S Extended Squitter (1090ES)

The Mode S Extended Squitter (1090ES) has been developed as an extension to the Mode S technology (described in Chapter 2) for Secondary Surveillance Radar (SSR). It supports Mode A/C radar, Mode S radar, MLAT, TCAS and ADS-B. The data link transmits at 1090 MHz. It is used widely in the United States and Australia for ADS-B Out service for commercial aircraft in non-radar airspace. European countries and China are using the data link for ADS-B Out operations in radar and non-radar airspace. It suffers from multipath (e.g. reflections from buildings, aircraft, etc.), also making it unsuitable for airport applications. There are also concerns about overloading at this frequency, which is, for example, also occupied by TCAS.

4.4.2.4 Data link performance

Data link performance is assessed in terms of the transmission range (air-to-air and air-to-ground), bandwidth (BW) and the vulnerability of the data link to external factors (interference, multipath, FRUIT, signal jamming). NATS and the EUROCONTROL Experimental Centre have conducted trials to compare the link performance of Mode S 1090ES, UAT and VDL4 on the Heathrow Airport surface for a selected set of trajectories and receiving ground station positions. The trial results (NATS, 2002) indicate that:

- During the static trial, messages were lost from the 1090ES when aircraft equipped with the same data link (1090ES) passed close to the test vehicle. The likely cause assumed is co-channel interference from the passing aircraft transponder rather than a temporary obstruction by the aircraft structure. The messages from UAT were not affected.
- 1090ES and UAT showed a reduction in the reception probability when the test vehicles were close to the Distance Measuring Equipment (DME). This is assumed due to:
  - corruption from the DME;
  - signal blocked by the obstruction; or
  - reflection or multipath due to the obstruction.

  However, VDL4 did not suffer from performance degradation.
- 1090ES and UAT were not available in certain sectors of the airport, where there was an obstruction between the ground stations and the test vehicle. VDL4 on the other hand was unaffected by the obstructions. This shows that the 1090ES and UAT signals were blocked due to the obstruction.
- 1090ES showed a lower performance with reception probability of 94.7 per cent compared to UAT at 99.9 per cent and VDL4 at 100 per cent. This is assumed to be due to other users of the same frequency (1090ES) on the airport surface. Further studies are required to confirm this hypothesis.
- UAT had better link reliability than 1090ES but suffered from line of sight problems. VDL4 had the highest link reliability.

4.5 ADS-B using Mode S 1090 MHz Extended Squitter (1090ES)

Mode S technology has two types of squitter, a short (56 bit) DF11 acquisition squitter and the extended (112 bit) DF17 squitter. The squitter is a reply format transmission without being interrogated by another means such as radar. The downlink format (DF) and uplink format (UF) are the two functional components of Mode S. UF is a specific interrogation originating from SSR or other aircraft requesting specific information from the aircraft. DF is the reply from the aircraft to the UF interrogation. The DF17 extended squitter is similar to elementary and enhanced surveillance (explained in Chapter 2) except that it does not need interrogation, i.e. it simply "broadcasts". The DF17 extended squitter includes the airborne position (BDS 0, 5), surface position (BDS 0, 6), extended squitter status (BDS 0, 7), identity and category (BDS 0, 8) as well as airborne velocity (BDS 0, 9). Binary Data Store (BDS) is a register within the transponder maintaining avionics data in 256 different 56 bit wide registers. It can be loaded with information and read out by the ground system. Each register contains the data payload of a particular Mode S reply or extended squitter. The BDS registers are also known as Ground Initiated Comm B (GICB) registers (ICAO, 2004c). The registers that are not updated within a fixed period are cleared by the transponder. Registers are identified by a two digit hex number. For example, BDS 05h (or also represented as BDS 0, 5) is the position squitter (Selex System Integration, 2013). In addition to the 56 bits, the Mode S short acquisition squitter includes:

- 8 bit CONTROL
- 24 bit ICAO aircraft address
- 24 bit PARITY.

1090ES includes an additional 56 bits data field used to carry ADS-B information. Table 4.7 presents the 1090ES ADS-B message type DF17, register and broadcast rate of each register. Figure 4.4 presents the data format.

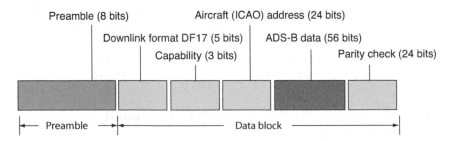

Figure 4.4 ADS-B Extended Squitter data format.
Source: modified from EUROCONTROL, 2007.

Table 4.7 1090ES ADS–B message, register and broadcast rates

Transponder register	Event driven message priority	1090ES ADS-B message	Broadcast rate		
			On the ground not moving	On the ground and moving	Airborne
BDS 0,5	N/A	Airborne position	N/A	N/A	2/1 second (0.4–0.6 sec)
BDS 0,6	N/A	Surface position	LOW RATE 1/5 seconds (4.8–5.2 sec)	HIGH RATE 2/1 second (0.4–06 sec)	N/A
BDS 0,8	N/A	Aircraft identification and category	LOW RATE 1/10 seconds (9.8–10.2 sec)	HIGH RATE 2/1 second (4.8–5.2 sec)	HIGH RATE 2/1 second (4.8–5.2 sec)
BDS 0,9	N/A	Airborne velocity	N/A	N/A	2/1 second (0.4–0.6 sec)
BDS 6,1	TCAS RA = 1 Emergency = 2	Aircraft status (Emergency/P (priority status), subtype = 1) (TCAS RA broadcast, subtype = 2)		TCAS RA or Mode A change 0.7–0.9 seconds No TCAS RA, no Mode A change 4.8–5.2 seconds	TCAS RA or Mode A change 0.7–0.9 seconds No TCAS RA, no Mode A change 4.8–5.2 seconds
BDS 6,2	N/A	Target State and Status (TSS)	N/A	N/A	1.2–1.3 seconds
BDS 6,5	N/A	Aircraft operational status	4.8–5.2 seconds	No change NIC_{SUPP}/NAC/SIL 2.4–2.6 seconds Change in NIC_{SUPP},NAC/SIL 0.7–0.9 seconds	TSS being broadcast or not No change TCAS/NAC/SIL/NIC_{SUPP} 2.4–2.6 seconds TSS being broadcast Change in TCAS/NAC/SIL/NIC_{SUPP} 2.4–2.6 seconds TSS not broadcast Change in TCAS/NAC/SIL/NIC_{SUPP} 0.7–0.9 seconds

Source: RTCA, 2011a.

The extended squitter illustrated in Figure 4.4 is composed of a pre-amble (8 bits) – also known as control bit – aircraft ICAO address (24 bits), parity check (24 bits), capability (3 bits), downlink format (5 bits) and ADS-B message (56 bits). The preamble bit is required to allow synchronization on reception. The parity check represents the error detection code with the capability bit indicating the capability of the Mode S transponder. The downlink format value is 17, representing the message type – ADS-B. The ADS-B message is defined in Table 4.6. It is also defined in Mode S Extended Squitter Standards and Recommended Practices (SARP) (Annex 10 Am. 77) (ICAO, 2002b) and Mode S Specific Services (ICAO, 2004c). The total duration of the extended squitter message is 120 μs (8 μs preamble and 112 μs data block). The data block is transmitted using Pulse Position Modulation (PPM). PPM is a relatively simple modulation scheme for a 1090 MHz receiver to decode in the presence of non-overlapping (in time) replies (Institute of Air Navigation Services, 2003).

The minimum content of an ADS-B message is composed of the following (De Oliveira *et al.*, 2009):

- Emitter Category – defining characteristics of the end users, for example light, medium or heavy aircraft, helicopters, UAV (Uninhabited Aerial Vehicle), land vehicles and obstacles;
- Emitter identifier – corresponding to the 24 bit network address in ATN;
- Latitude, longitude, flight level – corresponding to the 3D position of emitter end user;
- Aircraft identification – corresponding to the aircraft identification code (Squawk code);
- Data quality indicators – describing the integrity and accuracy of the data.

In addition to the above parameters, the ADS-B message also contains velocity, time stamp and intent information (in the latest version based on DO-260B (RTCA, 2011a)).

4.5.1 ADS-B position encoding and decoding – Compact Position Reporting (CPR) algorithm

ADS-B position data are provided in the World Geodetic System (WGS-84) format, latitude and longitude. Compact Position Reporting (CPR) was developed for ADS-B messages broadcast on the 1090ES

Extended Squitter (ES) datalink to reduce the number of bits required to transmit the latitude and longitude information. Position resolution for ES messages is approximately 5.1 metres for an airborne target and 1.3 metres for a surface target (Sensis Corporation, 2009). The circumference of the earth is approximately 40 000 kilometres so 40 000 000 m/5.1 m = ~7 800 000 discrete position values. 7 800 000 position values would require 23 bits in a message. Longitude is expressed over a range of 360° so longitude would require the full 23 bits. Latitude is expressed over a range of 180° so only 3 900 000 discrete position values or 22 bits would be required. Similarly, surface position would require 25 bits for longitude and 24 bits for latitude. CPR transmits position with 17 bits each for latitude and longitude plus 1 "CPR format" bit. Table 4.8 tabulates the message bits required for ADS-B position encoding with and without CPR for airborne and surface targets.

CPR saves 10 bits per position message for airborne targets and 14 bits per position message for surface targets. Position messages are envisioned to broadcast twice per second under most conditions. Therefore, CPR saves 20 bits/second for airborne targets and 28 bits/second for surface targets (Sensis Corporation, 2009).

4.6 System architecture

The optimum ADS-B architecture is unknown because it depends on the type of the data links planned to be used by the regions. As

Table 4.8 Message bits required for position encoding with and without CPR (Sensis Corporation, 2009)

		Without CPR	With CPR	Bits saved with CPR
Airborne position	Latitude	22	17	
	Longitude	23	17	
	CPR format	0	1	
	Total	45	35	10
Surface position	Latitude	24	17	
	Longitude	25	17	
	CPR format	0	1	
	Total	49	35	14

Source: Sensis Corporation, 2009.

discussed in section 4.4.2, ICAO proposes 1090ES as the global data link for ADS-B while UAT and VDL4 will be used at the regional level. The ADS-B system analysed in this book, and hence the architecture, is based on 1090ES. ICAO enumerated a number of functional requirements related to various surveillance applications without stipulating those to be supported by ADS-B (ICAO, 2003a). The surveillance applications envisioned to be supported by the ADS-B system are described in section 4.7. Figure 4.5 depicts the high level architecture of a complete ADS-B system.

The ADS-B system depicted in Figure 4.5 is composed of a message generation unit (which merges the data coming from aircraft sensors such as the navigation sensor, barometric altimeter and pilot inputs including aircraft identification and flight intent information), a transmitter (for transmission of the message), a data link (which carries out message distribution), a receiver (that receives the message) and a message processing unit (this prepares the ADS-B report for the use of various surveillance applications as described in section 4.6.2). Thus, unlike radar surveillance technology, which does not require a means of communication, ADS-B incorporates communication requirements to deliver the surveillance functions. In addition, ADS-B relies on the on-board navigation equipment to obtain aircraft positioning information. This shows a strong dependency between surveillance, navigation and communication functions

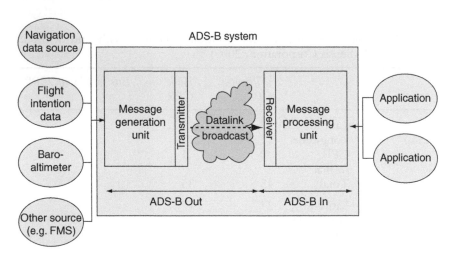

Figure 4.5 High-level architecture of ADS-B.

Source: ICAO, 2003a.

in ADS-B technology. Therefore, it is important to apply the Required Communication Performance (RCP), Required Surveillance Performance (RSP) and Performance Based Navigation (PBN) requirements to ADS-B system design and its implementation. The ADS-B system is integrated with many other external systems such as GNSS, which provides the navigation data; barometric altimeter, which provides the aircraft altitude; a module that allows the pilot to manually key-in messages such as flight intent or aircraft identification; and others. The application functions linked to the ADS-B system in Figure 4.5 represent the surveillance applications that utilize the ADS-B data to provide either air-air surveillance such as Cockpit Display of Traffic Information (CDTI), Aircraft Separation Assurance System (ASAS) or air-ground surveillance for the ATC.

4.6.1 System integration

The ADS-B system is a complex system, being highly dependent upon navigation and communication technologies. The ADS-B system is integrated with the following sub-systems to generate the ADS-B report, for broadcast to the ground-based ATC and to other ADS-B equipped aircraft within its configured range:

- navigation system
- barometric altimeter
- pilot input module (FMS/control panel)
- data link medium
- transmitter
- receiver.

The data generated by the individual avionic systems listed above are integrated in the ADS-B emitter/ADS-B capable transponder into a report which is broadcast to ground stations and other ADS-B equipped aircraft within the coverage area. Figure 4.6 shows a context diagram illustrating the data sources, elements and flow for ADS-B.

4.6.2 ADS-B report generation

An ADS-B message is a block of formatted data which composes an ADS-B report in accordance with the properties of the ADS-B data link (RTCA, 2002). The data link determines the size and type of

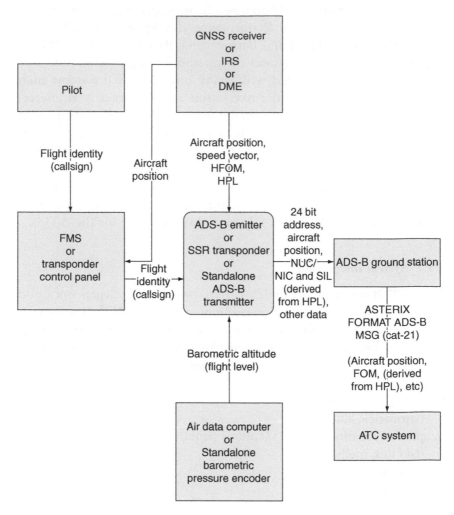

Figure 4.6 Context diagram for ADS–B data source, element and flow.

information that can be broadcast. ADS–B reports are specific information provided by the ADS–B Report Assembly Function to external applications supported by ADS–B. The report contains identification, state vector and status/intent information. The elements of the ADS–B report used and the frequency with which they must be updated vary by application. The portions of an ADS–B report that are provided vary by the capabilities of the transmitting ADS–B system. Figure 4.7 illustrates the report generation process and the corresponding modules.

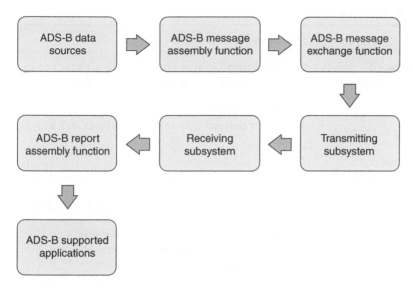

Figure 4.7 ADS–B report generation process.

4.7 Supported applications

ADS–B supports two types of applications:

- aircraft-to-aircraft applications, i.e. applications that transmit data from one aircraft or vehicle to others in the air and on the ground; and
- aircraft-to-ground applications, i.e. applications that require data to be broadcast from an aircraft or vehicle to fixed ground users (RTCA, 2002).

These applications can be categorized into three groups as shown in Table 4.9. The ADS–B data elements required to support enhanced air navigation and surveillance applications are summarized in Table 4.10.

4.8 ADS–B performance parameters, indicators, requirements and standards

ADS–B performance is measured in terms of accuracy, availability, integrity, continuity and latency of the surveillance data provided by the system (EUROCONTROL, 2011a). These parameters are defined in the following sub-sections. The quality indicators representing the surveillance data accuracy and integrity are derived from the on-board navigation source that feeds the ADS–B system with aircraft position

Table 4.9 ADS-B supported applications

Category	Application
Ground-based surveillance applications	1 ATC surveillance in airspace with radar coverage 2 ATC surveillance in airspace without radar coverage 3 Airport surface surveillance 4 Aircraft derived data for ground-based ATM tools
Aircraft-based surveillance applications	1 Situational awareness • Enhanced traffic situational awareness on the airport surface • Enhanced traffic situational awareness during flight operations • Enhanced visual acquisition of traffic • Enhanced successive visual approaches 2 Airborne spacing and separation • Enhanced sequencing and merging operations • In-trail procedure • Enhanced crossing and passing operations
Other applications	1 Ramp control/gate management 2 Noise monitoring 3 Flight following (for flying schools) 4 Remote airport charges issuing 5 Enhanced situational awareness of obstacles 6 Search and Rescue (SAR)

and velocity. Therefore, the ADS-B surveillance data performance mainly is driven by the on-board navigation system. In addition, the surveillance data performance also relies on the performance of the communication system that broadcasts the ADS-B surveillance data to users. Furthermore, various other factors affect the surveillance data performance. ICAO in collaboration with the RTCA, FAA, EURO-CONTORL and Air Services Australia has developed various requirements and standards to ensure adequate system performance and interoperability. The requirements and standards discussed in this section only apply to ADS-B Out. To date, the complete standards or requirements for ADS-B In are still to be developed.

4.8.1 *ADS-B performance parameters*

ADS-B performance is measured in terms of accuracy, availability, integrity, continuity and latency.

Table 4.10 ADS–B information required to support ADS–B applications

Information element	Aid to visual acquisition	Conflict avoidance and collision avoidance	Separation assurance and sequencing	Flight path deconfliction planning	Simultaneous approaches	Airport surface (A/V to A/V and A/V to ATS)	ATS surveillance
Identification							
Callsign	✓		✓	✓	✓	✓	✓
Address		✓	✓	✓	✓	✓	✓
Category			✓	✓	✓	✓	✓
State vector							
Horizontal position	✓	✓	✓	✓	✓	✓	✓
Vertical position	✓	✓	✓	✓	✓		✓
Horizontal velocity	✓	✓	✓	✓	✓	✓	✓
Vertical velocity	✓	✓	✓	✓	✓		✓
Heading		✓	✓	✓	✓	✓	✓
NIC						✓	✓
Mode status							
Emergency/priority status							✓
Capability codes		✓	✓	✓	✓	✓	✓
Operational modes		✓	✓	✓	✓	✓	✓
State vector quality		✓	✓	✓	✓	✓	✓
Air-reference vector		✓	✓	✓	✓		✓
Intent		✓		✓			✓

Source: RTCA, 2002.

4.8.1.1 ADS-B accuracy

ADS-B accuracy is defined as a measure of the difference between the aircraft position reported in the ADS-B message field and the true position. It is also defined as noise where the noise is assumed to follow a Gaussian distribution and the RMS value is quoted (ICAO, 2006b). ADS-B accuracy is also analysed based on the quality indicator representing position estimate accuracy included in the ADS-B message. The quality indicator derivation and definition are described in the next sub-section.

Horizontal position accuracy is assessed as the horizontal position measurement error distribution. For ADS-B, horizontal position accuracy is defined as the radius of a circle centred on the reported position of the target such that the probability of the actual position of the target being inside the circle is 95 per cent (ICAO, 2006b). This is illustrated in Figure 4.13.

Vertical accuracy is defined as the vertical position measurement error distribution. For ADS-B, a barometric altimeter on the aircraft provides the altitude to the ADS-B emitter and transmitted to the ADS-B ground station (ICAO, 2006b). In addition, ADS-B also provides geometric altitude derived by the on-board navigation system. However, the altitude data from the barometric altimeter is the current standard requirement for ATC operations even though the geometric altitude provides greater accuracy. Therefore, accuracy of the vertical position can be measured with reference to the geometric altitude.

Contributing elements to ADS-B accuracy include accuracy of the on-board navigation function that provides the positioning data to the ADS-B system, on-board latency and delay in the ADS-B communication function.

4.8.1.2 ADS-B integrity

ADS-B integrity is the level of trust that errors will be correctly detected. Integrity risk is the probability that an error larger than a given threshold in the information is undetected for longer than a predefined time to alert (ICAO, 2006b). ADS-B horizontal position integrity is the level of trust that can be placed in the navigation source to provide the input to the ADS-B reported position. This is represented by the integrity quality indicator derived from the navigation source position integrity indicator. The derivation of the integrity quality indicator included in the ADS-B message is explained in section 4.8.2. ADS-B position integrity is also analysed based on the quality indicator.

4.8.1.3 ADS-B continuity

ADS-B continuity is the probability that the system performs its required function without unscheduled interruption, assuming that the system is available when the procedure is initiated (ICAO, 2006b). ADS-B continuity includes:

*   the continuity of functions affecting all aircraft (e.g. satellite function, ground data acquisition function): expressed in terms of number of disruptions per year;
*   the continuity of system affecting only one aircraft (e.g. transponder function): expressed per flight hour; and
*   the continuity of navigation sources (including satellite constellations) of sufficient quality in the region, which affects many aircraft.

4.8.1.4 ADS-B availability

ADS-B availability is the ability of the system to perform its required function at the initiation of the intended operation. Availability is measured by quantifying the proportion of time the system is available with respect to the time the system is planned to be available. Periods of planned maintenance are not included in the availability measure (ICAO, 2006b). ADS-B availability includes:

*   the availability of functions affecting all aircraft (e.g. external positioning function, ground data acquisition function);
*   the availability of systems affecting only one aircraft (e.g. transponder function): expressed per flight hour; and
*   the availability of navigation sources (including satellite constellations) of sufficient quality in the region will affect many aircraft.

4.8.1.5 ADS-B latency

ADS-B latency is the delay between the aircraft position determination by the on-board navigation system and the position reception by the ground station. The latency measure directly affects the position accuracy.

4.8.2 Generation of surveillance data performance indicators

The ADS-B system obtains aircraft horizontal position in the World Geodetic System-84 (WGS-84) coordinates from the on-board GPS. The ADS-B reports delivered to ATC contain indicators of the position accuracy (Navigation Accuracy Category for Position – NACp) and integrity (Navigation Integrity Category – NIC). These indicators are based on the GPS integrity monitoring capability (RAIM) which reports the Horizontal Protection Level (HPL) with a 10^{-7}/hr integrity risk (encoded as NIC) and Horizontal Figure of Merit (HFOM) as a 95 per cent horizontal accuracy bound (encoded as NACp). HFOM is also known as Estimated Position Uncertainty (EPU).

The NIC parameter specifies a position integrity containment radius (Rc). NIC is reported such that ATC or other aircraft may determine whether the reported geometric position has an acceptable level of integrity for the intended use (Federal Aviation Administration, 2010a). Table 4.11 tabulates the applicable NIC values.

The NACp specifies the accuracy of the aircraft's horizontal position information (latitude and longitude) transmitted from the aircraft's avionics. Table 4.12 provides the applicable NACp values.

The block diagram in Figure 4.8 illustrates the interface and data flow between a GPS receiver and ADS-B system on board.

Table 4.11 NIC values

NIC	Containment radius
0	Unknown
1	RC < 37.04 km (20 NM)
2	RC < 14.816 km (8 NM)
3	RC < 7.408 km (4 NM)
4	RC < 3.704 km (2 NM)
5	RC < 1852 m (1 NM)
6	RC < 1111.2 m (0.6 NM)
	RC < 926 m (0.5 NM)
	RC < 555.6 m (0.3 NM)
7	RC < 370.4 m (0.2 NM)
8	RC < 185.2 m (0.1 NM)
9	RC < 75 m
10	RC < 25 m
11	RC < 7.5 m

Source: Federal Aviation Administration, 2010a.

Table 4.12 NACp values

NACp	Horizontal accuracy bound
0	EPU ≥ 18.52 km (10 NM)
1	EPU < 18.52 km (10 NM)
2	EPU < 7.408 km (4 NM)
3	EPU < 3.704 km (2 NM)
4	EPU < 1852 m (1 NM)
5	EPU < 926 m (0.5 NM)
6	EPU < 555.6 m (0.3 NM)
7	EPU < 185.2 m (0.1 NM)
8	EPU < 92.6 m (.05 NM)
9	EPU < 30 m
10	EPU < 10 m
11	EPU < 3 m

Source: Federal Aviation Administration, 2010a.

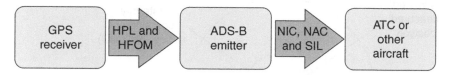

Figure 4.8 Data flow between navigation source and ADS-B equipment.

In addition to the NACp and NIC, ADS-B system performance is determined by the Source Integrity Level (SIL) parameter (Smith *et al.*, 2006). SIL is defined as the probability of the integrity containment radius used in the NIC parameter being exceeded without detection (ICAO, 2006b). The GPS HPL is encoded as the NIC at a SIL corresponding to 10^{-7} per hour, which is equivalent to SIL=3. ADS-B surveillance safety is assured by the NIC/SIL integrity parameters (ICAO, 2006b). Figure 4.9 illustrates the coded performance parameter for ADS-B based on GPS as a navigation source.

4.8.3 ADS-B performance requirements

Performance requirements for surveillance are determined by the application, including the airspace in which the aircraft operates. For example, reduced separation minima for the terminal area require better performance than in the en-route sector. The general requirements for the performance stipulated in the SPI-IR (EUROCONTROL, 2011a) and ED-142 (EUROCAE, 2010) are summarized in Table 4.13.

Table 4.13 Summary of ADS-B surveillance performance requirements

Item	Integrity	Accuracy	Continuity	Latency
Overall ADS-B system	$\leq 10^{-5}$ per flight hour (with respect to NIC) with time to alert ≤ 10 seconds	<150 metres for 3 NM separation	Update rate of ≤ 2 seconds	Total latency ≤ 1.5 seconds in 95% of transmissions. Uncompensated latency ≤ 0.6 seconds in 95% of transmissions. Uncompensated latency ≤ 1.0 second in 99% of transmissions

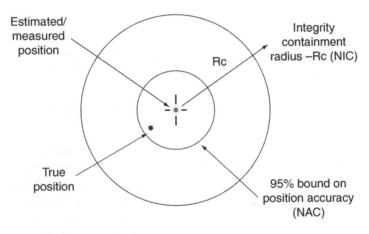

Figure 4.9 Coded performance parameters for ADS-B.
Source: modified from ICAO, 2006b.

4.8.4 ADS-B performance standards

ADS-B standards for avionic and ground station equipment are discussed in section 4.4. In this section, ADS-B performance standards are discussed. The main standards are developed by the RTCA and used globally by regulators, ANSPs, airline operators and aviation equipment manufacturers. The standards include ADS-B performance standards for operations, system and safety and interoperability.

4.8.4.1 Operational performance standard

The operational performance standard is provided in the "Minimum Operational Performance Standards (MOPS) for the 1090 MHz Extended Squitter Automatic Dependent Surveillance Broadcast (ADS-B) and Traffic Information Services Broadcast (TIS-B)", also known as RTCA DO-260B. This revision supersedes DO-260 and DO-260A. This document contains the minimum operational performance for airborne equipment for Automatic Dependent Surveillance-Broadcast (ADS-B) and Traffic Information Service-Broadcast (TIS-B) utilizing the 1090 MHz Mode S Extended Squitter (1090ES). Compliance with these standards by manufacturers, installers and users is recommended as one means of assuring that the equipment will satisfactorily perform its intended functions under conditions encountered in routine aeronautical operations. To date most of the operational aircraft flying were certified under the DO-260 standard. The equipage upgrade specifically for aircraft only took place at the beginning of 2015. This is discussed in section 4.4. The technical differences between DO-260, DO-260A and DO-260B were analysed by the ADS-B Study and Implementation Task Force (ICAO, 2012a). The findings are summarized in Table 4.14.

The improvements made from RTCA DO-260 to DO-260B resulted in additional information. This information is meant to increase the user confidence on the ADS-B information and as a door to enable the development of further ATM automation application for enhanced surveillance functions. However, it is important to note that these improvements in the standards do not improve the individual performance parameters. The only improvement noted will be due to the change in SDA (proposing direct connection between the on-board navigation system and ADS-B transponder) which will reduce latency of the ADS-B message, contrary to the system architecture proposed in the earlier standard.

The new information available under DO-260A/B would benefit the users in terms of operational decisions as follows (ICAO, 2012a):

• The additional quantum levels of NIC would provide the ANSPs more flexibility in deciding whether the NIC is considered as "good". For example, if it is decided that $Rc < 0.6$ NM can be used for radar separation, instead of $Rc < 0.5$ NM, more aircraft could benefit from the ADS-B services. Instead in DO-260, the level immediately after HPL < 0.5 NM is HPL < 1.0 NM.

Table 4.14 Differences between DO–260, DO–260A and DO–260B

	DO-260	DO-260A	DO-260B	Availability of data in ASTERIX CAT 21
Introduction of Navigation Integrity Category (NIC) to replace Navigation Uncertainty Category (NUC$_P$)	NUC$_P$ is used	NIC is used to replace NUC$_P$	More level of NIC available. Vertical component removed	NIC is shown in v1.0 and above. More level of NIC (shown as PIC) is available in v2.1
Quality Indicator for Velocity (NUC$_R$ and NAC$_V$)	NUC$_R$ is used	Replaced with NAC$_V$. Definition remains the same	Vertical component removed	Available in v0.23 and above
Surveillance Integrity Level and Source Integrity Level (SIL)	Not available	Surveillance Integrity Level is used	Renamed as Source Integrity Level. Definition is changed to exclude avionics fault	Available in v1.0 and above
System Design Assurance (SDA)	Not available	Not available	To address probability of avionics fault	Available in v2.1
Navigation Accuracy Category (NAC$_P$)	Not available	Derived from HFOM and VFOM	Relies only on HFOM	Available in v1.0 and above
Geometric Vertical Accuracy (GVA)	Not available	Not available	Derived from VFOM	Available in v2.1
Barometric Altitude Integrity Code (NIC$_{BARO}$)	Not available	To indicate integrity of barometric altitude	Same as DO-260A	Available in v1.0 and above
Length/width of Aircraft	Not available	Provides an indication of aircraft size	Same as DO-260A	Available in v1.0 and above

Indication of capabilities	Only shows status of TCAS and CDTI	More information available including capability to send Air Reference Velocity, Target State and Trajectory Change reports	Additional information on type of ADS-B In (i.e. 1090ES In or UAT In)	Available in v1.0 and above, except availability of 1090ES/UAT In and information on GPS antenna offset
Status of Resolution Advisory		Information on whether Resolution Advisory is active	Same as DO-260A	Available in v1.0 and above
GPS offset		Indication on whether GPS offset is applied	Information on GPS antenna offset is provided	GPS offset status is available in v1.0 and above. Information on GPS offset is not available in ASTERIX
Intention	Not available	Able to indicate intended altitude and heading	Same as DO-260A	Intended altitude is available in v0.23. Intended heading is not available in ASTERIX
Target status	Not available	Not available	Indication of Autopilot Mode, Vertical Navigation Mode, Altitude Hold Mode, Approach Mode and LNAV Mode	
Resolution Advisory	Not available	Not available	Availability of Active Resolution Advisories; Resolution Advisory complement record, Resolution Terminated; Multiple Threat encounter; Threat Type indicator; and Threat Identity data	Available in v1.0 and above
Mode A	Broadcasted using test message in USA only	Broadcasted using test message in USA only	Broadcasted worldwide as a regular message	Available in v0.26 and above

- The SIL will allow the user to further assess the integrity of the reported position.
- The SDA will indicate the robustness of the system, and allow the ANSPs to decide on a minimum SDA for ADS-B services.
- The NIC_{BARO} which indicates the integrity of the barometric height may potentially be used to develop new ATM tools in this feature.
- The width/length which indicates the size of the aircraft may be used as an input for generating alerts on airport surface movement control.
- Indication on GPS offset may be one of the inputs for generating alerts on airport surface movement control. Indication on the availability of 1090ES/UAT will allow the controller to anticipate a potential request for In-Trail Procedure clearance. Indication of the resolution advisory status allows the controller to know whether the pilots were alerted about the potential conflict.
- The intent heading and flight level can be used as an input to the trajectory prediction algorithm in the Short-Term Conflict Alert.
- The target status allows the controller to know the mode that the aircraft is in.
- The Resolution Advisory will help the controller know the advisories that are provided to the pilots by the ACAS. This will prevent the controller from giving instructions that are in conflict with the ACAS.
- The Mode A allows flight plans to be coupled with the ADS-B tracks.

4.8.4.2 System performance standard

The system performance standard is provided in the "Minimum Aviation System Performance Standards (MASPS) for Automatic Dependant Surveillance Broadcast (ADS-B)", also called RTCA DO-242A. This document supersedes DO-242 and provides an up-to date view of the system-wide operational use of ADS-B. This revised ADS-B MASPS concentrates on four major areas of development:

- separating the accuracy and integrity components of the Navigation Uncertainty Category (NUC) into the new fields Navigation Accuracy (NAC) and Navigation Integrity Category (NIC);
- reorganization of the State Vector, Mode-Status and On-condition reports;

- restructuring the content and manner in which intent information is broadcast; and
- clarification that system requirements at the MASPS level are based on operational ranges and not particular applications.

4.8.4.3 Safety and interoperability performance standard

The safety and interoperability performance standard is provided separately for non-radar and radar airspace in Safety, Performance and Interoperability Requirements for the ADS-B Non-Radar-Airspace (NRA) Application (RTCA DO-303) and "Safety, Performance and Interoperability Requirements for Enhanced Air Traffic Services in Radar-Controlled Areas Using ADS-B Surveillance (ADS-B-RAD)" (RTCA DO-318). These documents provide the requirements for ADS-B Out safety and interoperability in non-radar and radar airspace respectively.

It is important to understand that the requirements stipulated in the various standards by RTCA, EUROCAE or ICAO are meant to affirm the required functions and performance level. However, the standards do not provide the methods/mechanisms to implement those requirements. They are totally dependent on the equipment manufacturers. Hence, there is no standardized method available on how to develop the mechanism to enable the required functions within the system. The willingness of the equipment manufacturers to invest in the research and development of the system functionalities are also influenced by the mandate of the ADS-B system. Various versions of the equipment may induce different problems due to the different methods developed by the different manufacturers.

4.9 How can ADS-B fit into the ATC system?

The objective of ATM is "to enable aircraft operation to meet their planned times of departure and arrival and adhere to their preferred flight profiles with minimum constraints, without compromising the agreed levels of safety" (EUROCONTROL, 2006). The current ATM system (ICAO, 2000) is illustrated in Figure 4.10. Due to the technological advancements, the ATM system has incorporated airborne components: airborne CNS and Aircraft Collision Avoidance System (ACAS). This section focuses on ATC and airborne ATM components. These components are inter-reliant to ensure the safe control, monitoring and management of the air traffic. Figure 4.11

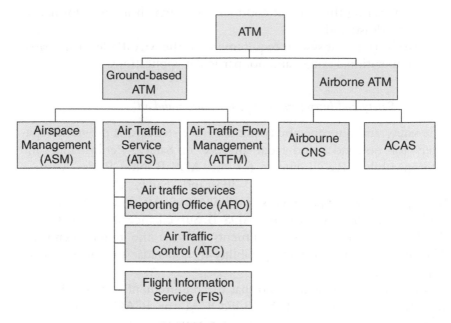

Figure 4.10 Air Traffic Management.
Source: ICAO, 2000.

shows the functional components in a new ATC paradigm (proposed in this book) combining ATC and Airborne Communication, Navigation and Surveillance (CNS).

4.9.1 New ATC system loop model

Hansman (1997) states that in the current ATC operation, the role separation between controllers and pilots is functional and ambiguous. Controllers have the responsibility for traffic separation while the pilot is responsible for the safety of the flight. The pilot will defer to controllers on the traffic issues as the controllers have more information on it. Likewise, controllers will often defer to pilots on weather information. Hence it is obvious that the functional role of pilots and controllers is directly determined by the availability of reliable information in hand. Bearing this in mind, the implementation of new surveillance technology such as ADS-B and TIS-B, and data link technologies, will enable both pilots and controllers to have access to the same situational information. Therefore, this may create conflict in the role separation between pilots and controllers. Emergence of

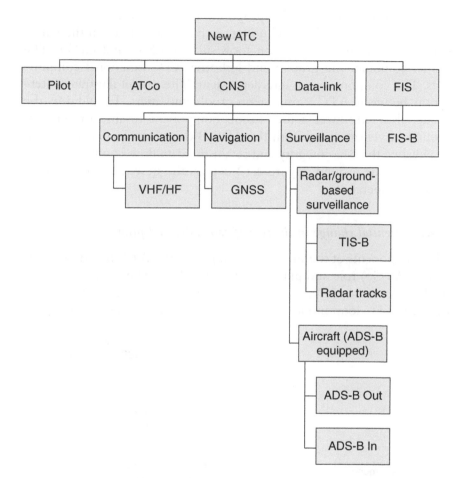

Figure 4.11 New ATC system component (proposed in this book).

the new technologies is expected to support new surveillance applications envisaged to balance both controller and pilot workload, optimize capacity, increase efficiency and improve safety. It is envisioned that the ADS-B system will be the primary surveillance source in the future. This will create controller dependency on the aircraft to obtain the surveillance data. Conversely, pilots will have less dependency on the controllers due to enhanced situational awareness. Radar surveillance and procedural control via communication aids will act as a backup system. Therefore, the ATC role structure has to be carefully designed and clearly enforced upon the implementation of the new technologies.

A new ATC loop model is proposed in this book as a result of the emergence of the ADS-B technology, which has resulted in the emergence of many new application tools such as ASAS and CDTI. The model combines the ground and airborne ATC functional components. The components are interdependent. The model illustrates interaction between ATC components, tools and users. In addition the model also introduces a data fusion component to fuse the current ground surveillance data with the ADS-B data. Data fusion is foreseen to enhance the data integrity and retain the data flow continuity in the air and on the ground. Figure 4.12 depicts the new ATC loop model for future operation.

4.9.2 Potential change in the role of controller and pilot

The components of the model are mapped to the ATC functions, controller (ATCo) role and pilot role in Table 4.15. The potential change in the role of the ATCo and pilot are analysed and mapped based on the new ATC loop model in Figure 4.12 for all phases of flight. Based

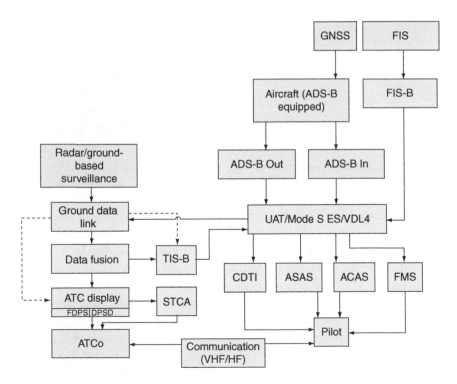

Figure 4.12 New ATC loop model.

Table 4.15 ATC paradigm shift based on phases of flight

ATC functions	Current role / Support technologies				New role / Support technologies			
	Airport surface	En-route terminal	En-route domestic	Oceanic/remote area	Airport surface	En-route terminal	En-route domestic	Oceanic/remote area
Conformance monitoring	ATCo — DPSD FDPS	ATCo — DPSD FDPS	ATCo — DPSD FDPS	ATCo — COM	ATCo — DPSD FDPS	ATCo — DPSD FDPS	ATCo — DPSD FDPS	ATCo — DPSD FDPS
Hazard monitoring	ATCo — DPSD FDPS COM	ATCo — DPSD	ATCo — DPSD	ATCo — FDPS COM	ATCo — DPSD	ATCo — DPSD	ATCo — DPSD	ATCo — DPSD
Sequencing	ATCo — DPSD FDPS COM	ATCo — DPSD FDPS COM	ATCo — DPSD FDPS COM	ATCo — FDPS COM	ATCo — DPSD FDPS COM	Pilot — IM CDTI ASEP ASAS COM	Pilot — IM CDTI ASEP ASAS	Pilot — IM CDTI ASEP ASAS
Spacing	ATCo — DPSD FDPS COM	ATCo — DPSD FDPS COM	ATCo — DPSD FDPS COM	ATCo — FDPS COM	ATCo/Pilot — DPSD FDPS COM IM	Pilot — IM CDTI ASEP	Pilot — IM CDTI ASEP	Pilot — IM CDTI ASEP
Merging	ATCo — DPSD FDPS COM	ATCo — DPSD FDPS COM	ATCo — DPSD FDPS COM	ATCo — FDPS COM	Pilot — IM CDTI	Pilot — IM CDTI	Pilot — IM CDTI	Pilot — IM CDTI
Conflict detection	ATCo — DPSD FDPS	ATCo — DPSD FDPS	ATCo — DPSD FDPS	Pilot — ACAS	ATCo — DPSD FDPS	Pilot — ACAS CDTI	Pilot — ACAS CDTI	Pilot — ACAS CDTI
Short-term conflict detection	ATCo — DPSD FDPS	ATCo — DPSD FDPS	ATCo — DPSD FDPS	Pilot — ACAS VISUAL	ATCo/Pilot — STCA ACAS CDTI	Pilot — ACAS CDTI	Pilot — ACAS CDTI	Pilot — ACAS CDTI
Conflict resolution and intervention	ATCo — COM	ATCo — COM	ATCo — COM	ATCo — COM	Pilot — SSEP ASAS	Pilot — SSEP ASAS ASEP	Pilot — SSEP ASAS ASEP	Pilot — SSEP ASAS ASEP
Flight replanning	ATCo — FDPS COM	Pilot — FMS COM	Pilot — FMS COM	Pilot — FMS COM	ATCo — FDPS COM	Pilot — FMS CDTI	Pilot — FMS CDTI	Pilot — FMS CDTI
Conformance to ATC vector	Pilot — COM	Pilot — COM	Pilot — COM	Pilot — COM	Pilot — ATSAW-SURF	Pilot — CDTI AIRB	Pilot — CDTI AIRB	Pilot — CDTI AIRB
On-board collision avoidance	Pilot — ACAS	Pilot — ACAS	Pilot — ACAS	Pilot — ACAS	Pilot — ACAS CDTI AIRB	Pilot — ACAS CDTI AIRB	Pilot — ACAS CDTI AIRB	Pilot — ACAS CDTI AIRB
Information feed and support	ATCo — COM	ATCo — COM	ATCo — COM	ATCo — COM	ATCo/Pilot — FIS-B TIS-B	ATCo/Pilot — FIS-B TIS-B	ATCo — FIS-B	ATCo — FIS-B

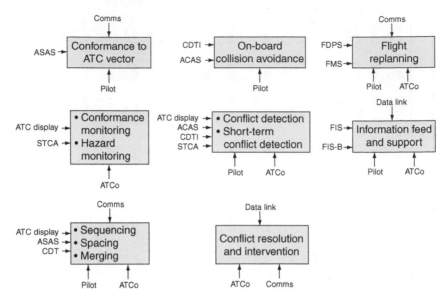

Figure 4.13 ATC functional blocks.

on the mapping, with the existence of ADS-B and its applications, the role of ATCo will be more focused on monitoring rather than controlling in the future, while pilots will be able to self navigate. Based on the mapping, further analysis is made using a widely used approach for functional modelling, Structured Analysis and Design Technique (SADT). The approach was introduced by Ross of Soft Tech Inc. in 1973 and is further described by Lissandre (1990) and Lambert *et al.* (1999). In the SADT diagram, each functional block is modelled with five main elements: functions, input, control, mechanism and output. Figure 4.13 illustrates the ATC functional blocks analysis with the new applications and technologies as input. The role shift in Table 4.15 and ATC functional blocks in Figure 4.13 provide an insight and pave the way to the future ATC operations with enhanced ground surveillance applications and airborne surveillance applications in place. The application input to each function in Figure 4.13 could be expanded as new applications emerge as a result of complete ADS-B implementation worldwide.

4.10 Summary

This chapter has introduced the ADS-B system along with detailed description of the requirements, system architecture including functions and how they are supported. The chapter proposed a new ATC loop model with the implementation of ADS-B. The model is important for ATC operations during the transition period and complete implementation of ADS-B. Finally, the chapter proposed a paradigm shift in the ATC operational tasks between the pilots and controllers, envisaged upon complete ADS-B implementation in the future.

5 Safety improvement potentials with ADS-B

5.1 Safety improvements

The capabilities of ADS-B are outlined in Chapter 4. High-performance ADS-B positioning information (in terms of accuracy, integrity and update rate) in real time is required to enable many ground and airborne applications in the future (Butterworth-Hayes, 2012). This can be verified by the functionalities supported by the airborne applications. For example, on the airport surface, a traffic display (CDTI) in the cockpit showing all aircraft taxiing on the manoeuvring area and within the terminal area, coupled with a background airport map from Flight Information Service Broadcast (FIS-B), would improve the pilot's situational awareness, reduce the risk of runway incursions and assist the pilot to navigate around an unfamiliar airport (AirservicesAustralia, 2007). Furthermore, it will improve safety during extreme weather and reduced visibility conditions (Herrera *et al.*, 2009). In the case of reduced or unavailability of radio communication between pilots and controllers, availability of traffic display (CDTI) and assistance to perform separation (ASAS) will aid safe aircraft navigation.

The use of In-Trail Procedure (ITP) in a Controller-Pilot Data Link Communication (CPDLC) environment is accepted and positively rated by the flight crew and controllers (EUROCONTROL, 2009a). ADS-B provides a safety measure for this application by providing traffic display (CDTI), as a complement to more specific ITP information to improve situational awareness in support of flight level changes. The CRISTAL-ITP Project (EUROCONTROL, 2009a) concluded that the quality of the ADS-B Out information from the reference aircraft in terms of update rate, accuracy and integrity as received was sufficient to support ITP flight trial. The data recorded from the trial also showed that the received ADS-B Out information was compliant with Airborne Traffic Situational Awareness–In-Trail Procedure Safety Performance Requirements (ATSA-ITP SPR) (RTCA, 2008). This will enable aircraft surveillance in oceanic and remote areas (which did not exist with the current surveillance system) resulting in increased safety for aircraft navigation within these areas.

Over the past 20 years, the threat of a mid-air collision occurring on a commercial flight has tremendously decreased and is very rare (Zwegers, 2010). This is primarily due to the implementation of TCAS. However, the TCAS system was not designed for small General Aviation (GA) aircraft due to its size and prohibitive cost (Zwegers, 2010). Therefore, mid-air collisions involving GA still occur, especially with student pilots on board.

ADS-B data can be used to improve TCAS surveillance and traffic display. This technique is known as hybrid surveillance (RTCA, 2006b). The use of ADS-B data (passive surveillance) enables TCAS traffic displays to more accurately depict the bearing and velocity of surrounding aircraft. In addition, the identity information received through ADS-B allows identification of other aircraft. Furthermore ADS-B enables TCAS to track aircraft at a range of more than 100 NM compared with 40 NM using TCAS active surveillance (Airservices-Australia, 2007). Hence ADS-B improves situational awareness and safety for the TCAS application. Despite these advancements, the collision avoidance function within TCAS remains unchanged with Hybrid Surveillance; ADS-B information is not used as input to calculate resolution advisories. It is envisaged that in the future ADS-B data will be used in the TCAS collision avoidance function to take advantage of its high performance (accuracy and update interval). This is supported by research conducted by the Embry-Riddle Aeronautical University (ERAU). ERAU conducted a case study with 100 training aircraft fully equipped with ADS-B in a dense airspace. The study (Zwegers, 2010) indicated that ADS-B has dramatically decreased the risk of mid-air collision in very congested airspace by:

- providing pilots with real-time traffic information with high accuracy, in which to implement conflict detection and resolution, especially below radar coverage (low altitude and ground information), thereby avoiding mid-air collision and runway incursion;
- providing pilots with graphical and textual weather information; and
- providing ANSPs with real-time information of aircraft location for planning purposes (spreading out aircraft to minimize congestion) and flight following (tracking).

Furthermore, recordings of ADS–B data can be used by the stake-holders (airline operators, ANSPs and regulators) to increase safety and efficiency practices (e.g. incident/accident investigation, pattern flow in/out airspace study, address noise complaint).

5.1.1 ADS-B for the future ATM system modernization and improvements

At the core of the future SESAR and NextGen ATM are advanced automation systems based on the ADS–B. These must progressively fulfil a number of functions, as exemplified in Figure 5.1. Note that this may not be an exhaustive and/or validated list of functions. These functions may evolve in the future.

The first step requires the aircraft to be equipped with ADS–B Out. The second step involves the implementation of ground surveillance applications for ATC including in non-radar airspace (NRA) (step 3), in radar (RAD) airspace (step 4) and on the airport surface (APT) (step 5). The implementation is conducted in sequence, based on the

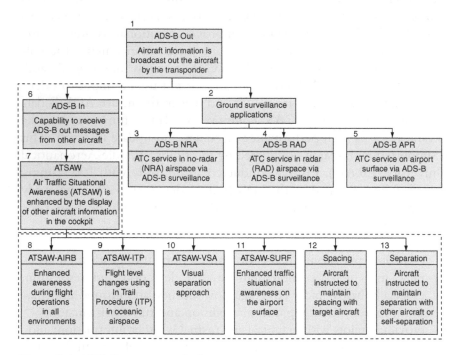

Figure 5.1 ADS–B system evolution.
Source: Ali *et al.*, 2016.

criticality of the limitations of the current radar system to support ATC to provide air traffic services to the aircraft. ADS-NRA has been fully implemented and is operational in various regions such as Australia while ADS-B RAD and ADS-B APT are still currently under trial in most regions. These applications are meant to provide radar like services where the radar is either unavailable, or to supplement the reduced radar services in a particular operational environment or airspace. Future applications envision providing enhanced surveillance services (e.g. reduced separation to aircraft) by exploiting the higher performance of ADS-B. However, these are still to be implemented due to a lack of confidence in the system performance and aircraft equipage.

The sixth step is the implementation of ADS-B In, which requires ADS-B In equipage to enable aircraft to receive ADS-B Out messages from other aircraft within their specified range. ADS-B In is a means to enable various airborne surveillance applications including providing ATSAW via the display of other aircraft information to the flight crew. At present, pilots build traffic situational awareness by integrating information from two main sources: visual observation and radio communication with ATC. The radio communication includes traffic information provided to flight crew by a controller, transmission from a controller to other aircraft, and responses from other aircraft, and air–to–air radio communication outside controlled airspace. Additionally, to enhance situational awareness, pilots with suitably equipped aircraft may use their TCAS traffic display to supplement the available traffic information. Even though the TCAS display is meant to support visual acquisition when the TCAS generates a Traffic Advisory (TA), in some cases it has confused the pilot's perception of the traffic situation (CASCADE Operational Focus Group, 2009). This causes unsynchronized situational awareness between pilots and ATC which may lead to undesirable incidents. According to EUROCONTROL (CASCADE Operational Focus Group, 2009), this particular situation has been one of the drivers of the development of airborne surveillance applications. ATSAW has led to the development of various surveillance applications: enhanced traffic situational awareness in all environments (ATSAW-AIRB), flight level changes using In-Trail Procedure (ITP) in oceanic airspace (ATSAW-ITP), visual separation approach (ATSAW-VSA) and enhanced traffic situational awareness on the airport surface (ATSAW-SURF). The successful implementation of the ground and airborne surveillance applications is underpinned by ADS-B Out performance that is sufficient for each of these applications. The next section describes the airborne applications, the operational environment and their requirements. This is followed, in the last section, by the performance of real time ADS-B being mapped to the requirements of these applications.

5.1.1.1 Airborne surveillance applications using ADS-B

This section reviews and discusses the various airborne surveillance applications envisioned with the ADS-B system, highlighted in Figure 5.1.

5.1.1.1.1 AIR TRAFFIC SITUATIONAL AWARENESS DURING FLIGHT OPERATIONS (ATSAW-AIRB)

The ATSAW-AIRB is defined as the enhancement of a flight crew's knowledge of the surrounding traffic situation in all environments. It is meant to improve flight safety and operations by assisting flight crews in building their traffic situational awareness through the provision of an appropriate on-board traffic display (CASCADE Operational Focus Group, 2009). This is achieved by retrieving ADS-B information transmitted by other aircraft transponders via Mode S 1090 MHz. The information is then fed to the Cockpit Display of Traffic Information (CDTI) tool to provide instantaneous and up-to-date traffic information (including aircraft identification, position, direction, ground speed, vertical tendency, relative altitude and wake vortex category). The use of ATSAW-AIRB does not require any changes to the ATS infrastructure, systems and ATC procedures (CASCADE Operational Focus Group, 2009). The ATSAW-AIRB application requires all aircraft within the airspace to be capable of transmitting ADS-B Out messages and the "owner aircraft" to be equipped with a traffic display (e.g. CDTI or merged TCAS/ADS-B traffic display). Standardization for the implementation of the application is developed jointly by EUROCAE and the RTCA (EUROCAE and RTCA, 2010). EUROCONTROL has developed a Preliminary Safety Case for ATSAW-AIRB. To date more than 3000 ATSAW-AIRB equipped flights have been performed in Europe (Rekkas, 2013). However, the relevant safety case is not publicly available.

5.1.1.1.2 AIR TRAFFIC SITUATIONAL AWARENESS IN-TRAIL
PROCEDURE IN OCEANIC AIRSPACE (ATSAW ITP)

Currently, aircraft operating in procedural airspace (oceanic or remote) are constrained to fly at the same flight level, and thus do not necessarily fly at an optimum flight level. ATSAW-ITP using ADS-B is meant to enable altitude changes. The ITP is achieved with the combination of ATSAW and Controller-Pilot Data Link Communication (CPDLC). The ATSAW display allows the pilot to detect a climb/descend opportunity. The clearance exchange for the altitude change is then requested via CPDLC. The shared situational awareness between pilot and the ATC enabled by ADS-B will provide confidence to ATC to grant the clearance requested. This will also lead to reduced separation between aircraft in these airspaces. The current standard longitudinal separation requirement is 80 NM (ICAO, 2007a), while with ATSAW-ITP, a reduced longitudinal separation of 15 NM (Vidal, 2012) is expected to be achieved.

5.1.1.1.3 AIR TRAFFIC SITUATIONAL AWARENESS VISUAL
SEPARATION IN APPROACH (ATSAW-VSA)

Visual separation is meant to separate aircraft (Instrument Flight Rules (IFR) and Visual Flight Rules (VFR)) by means of pilots seeing and avoiding other aircraft or by means of a tower controller directly observing and separating aircraft visually. ATSAW-VSA is meant to assist this type of operation for pilots. The objective of this application is to safely execute approach procedures using "own separation" from the preceding aircraft more efficiently and more regularly (CASCADE Operational Focus Group, 2008). It aids the pilot to acquire and maintain visual contact with the preceding aircraft. More importantly it supports safe operations in marginal meteorological conditions. The ATSAW-VSA improves efficiency by increasing the runway capacity, and also improves safety by providing enhanced situational awareness and enhanced identification of the target aircraft (Vidal, 2010). The ATSAW-VSA paves the way for future spacing applications. To enable the ATSAW-VSA, the aircraft has to be equipped with ADS-B In equipment, appropriate flight deck tools, and a traffic display tool (e.g. CDTI). Most importantly the application is only feasible with the full mandate of ADS-Out, ensuring all surrounding aircraft are equipped with ADS-B Out capability. Partial equipage of surrounding aircraft is not sufficient to use the ATSAW-VSA application.

5.1.1.1.4 AIR TRAFFIC SITUATIONAL AWARENESS ON THE AIRPORT SURFACE (ATSAW-SURF)

The ATSAW-SURF is intended to improve situational awareness of surrounding aircraft and ground vehicles operating in the vicinity of the aerodrome. This is achieved by providing the pilot with a display of the surrounding traffic position and identity, together with the "own aircraft" position overlaid on a map of the aerodrome. The enhanced situational awareness provided by the ATSAW-SURF application will improve the safety of aerodrome surface operations, in particular at taxiway and runway intersections, and for aircraft landing and taking off. A secondary outcome is to enhance taxi efficiency through improved traffic situational awareness during operations such as conditional taxi clearances, especially during low visibility conditions, night operations and at airports unfamiliar to flight crews. The application is also expected to decrease pilot and controller workload by reducing requests for repeat information with respect to surrounding traffic (ICAO, 2012b). To enable the ATSAW-SURF application, the aircraft has to be equipped with ADS-B In equipment, a traffic display tool and must have access to the airport map database.

5.1.1.1.5 SPACING/INTERVAL MANAGEMENT (IM)

The step following the introduction of ATSAW applications is the introduction of spacing applications (Vidal, 2012). This is also known as Interval Management (IM). According to the ICAO, IM provides improved means for managing traffic flows and aircraft spacing. This includes the use of both ground and airborne tools as follows (ICAO, 2012b):

- Ground tools that assist the controller in evaluating the traffic scenario and determining appropriate clearances to merge and space aircraft efficiently and safely, and allow the controller to issue an IM clearance; and
- Airborne tools that allow the pilot to conform to the IM clearance. These airborne capabilities are referred to as the Flight-deck based Interval Management (FIM) capabilities. The requirements for the FIM are provided in "Safety Performance and Interoperability Requirements for Flight Deck Interval Management" (EUROCAE, 2011).

Under the IM, the equipped aircraft is instructed to merge behind and maintain a given time spacing from another aircraft. Three types of manoeuvers are supported by the IM application:

- remain in trail;
- merge in trail; and
- radar vector then merge in trail.

This is illustrated in Figure 5.2. Compared with current operations, the controller is relieved of the provision of speed and turn clearance to manage traffic by assigning an interval to the pilot. However, during the IM operations, the controller still retains responsibility for separation.

IM is currently one of the core capabilities under NextGen and avionic standards to support the application have recently been published. To boot, numerous flight tests have been conducted in the National Airspace System (NAS). Furthermore, advanced IM capabilities are under research and developments to provide enhance operations and performance. Barmore *et al.* (2016) provide more information on the IM development and deployment plan in the NAS.

5.1.1.1.6 AIRBORNE SEPARATION (ASEP) AND AIRBORNE SELF-SEPARATION (SSEP)

The separation application refers to Airborne SEParation (ASEP) and Airborne Self-SEParation (SSEP). According to the ICAO (2012b), delegation of separation responsibility to flight crew is foreseen in the

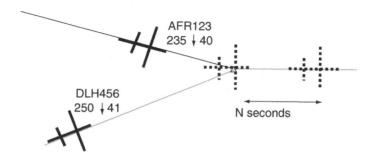

Figure 5.2 Manoeuvres supported by the Interval Management application.
Source: Ali *et al.*, 2016.

future. The pilot will be responsible to ensure separation from designated aircraft as communicated in future clearance, thereby relieving the controller from the responsibility for separation between these aircraft. Typical ASEP include (ICAO, 2012c):

- interval management with delegation of separation: the flight crew maintains a time-based separation behind designated aircraft;
- lateral crossing and passing: the flight crew adjusts the lateral flight path to ensure that horizontal separation with designated aircraft is larger than the applicable airborne separation minimum;
- vertical crossing: the flight crew adjusts the vertical flight path to ensure that vertical separation with designated aircraft is larger than the applicable airborne separation minimum;
- paired approaches in which the flight crew maintains separation on final approach to parallel runways; and
- in oceanic airspace, improved procedures of ITP using new airborne separation minima: ASEP-ITF In-trail follow; ASEP-ITP In-trail procedure; and ASEP-ITM In-trail merge.

During SSEP, the pilot ensures separation of their aircraft from all surrounding traffic. Hence the controller has no responsibility for separation. Typical airborne self-separation applications include (ICAO, 2012c):

- airborne self-separation in ATC-controlled airspace;
- airborne self-separation in segregated en-route airspace;
- airborne self-separation in mixed-equipage en-route airspace; and
- airborne self-separation – free flight on an oceanic track.

An early implementation of the ASEP and SSEP applications is anticipated in oceanic and low-density airspace. Advanced Safe Separation Technologies and Algorithms (ASSTAR) initiated the work on ASEP and SSEP applications in Europe which has been supported by two dedicated SESAR projects (EUROCONTROL, 2016b), i.e. 04.07.04.b ASAS-ASEP Oceanic Applications and 04.07.06 En Route Trajectory and Separation Management – ASAS Separation (Cooperative Separation).

Airborne separation minima have yet to be defined for the ASEP and SSEP applications. These are expected to be very stringent, leading to the requirement of very high performance navigation and surveillance functions on board. In addition, due to the impact of these applications on the controller and pilot responsibilities, provisions for these applications are expected to require modification of the ICAO annexes, i.e. PAN ATM (ICAO, 2007a) and PAN OPS (ICAO, 2006d).

All of the airborne surveillance applications discussed above rely on the capability of ADS-B Out to provide the required information elements with a specific level of performance. Tables 5.1 and 5.2 provide the required ADS-B information elements and the corresponding minimum ADS-B system performance requirements to support these applications respectively.

5.1.1.2 Space-based ADS-B for surveillance in oceanic, remote and polar areas

As discussed earlier in Chapters 2 and 3, radar systems are costly and not suitable to be deployed over large geographic areas (e.g. oceans) and limited to line-of-sight. At present ATC services in such areas are provided using Procedural Control (explained in Chapter 2) and Automatic Dependent Surveillance – Contract (ADS-C) for limited numbers of aircraft subscribing to the service. ADS-C is an extension of ADS technology whereby ADS data are transmitted to the stakeholders (e.g. airline operator or ANSP) via Controller-Pilot Data Link Communication (CPDLC), a data link application that relies on High Frequency Data Link (HF DL), VHF Data Link (VDL) and satellite communication (SATCOM). Even though ADS-C can enhance situational awareness for air traffic controllers in areas without surveillance coverage, the high cost of aircraft equipage typically restricts ADS-C to large, twin-aisle aircraft, with position updates only provided every 10–18 minutes (Aireon, 2016a).

Advancements in the satellite technology have enabled extension of ADS-B coverage over oceans, mountains, remote areas and polar regions to provide real-time surveillance of ADS-B equipped aircraft anywhere on the globe. This system is known as space-based ADS-B system, illustrated in Figure 5.3.

In addition, the space-based ADS-B surveillance will also overcome the limitations of ground-based radar, Wide Area Multilateration (WAM) and ADS-B Out surveillance systems, which are often constrained by geographical surface, cost, blind spots, delays and power requirements. Hence, leaving an estimated 70 per cent of global airspace without real-time aircraft surveillance coverage (Aireon, 2016a). A recent incident on 6 August 2016, Air Algerie flight AH1020 flying

Table 5.1 Required information elements to support ADS-B applications

Information element	ATSAW-AIRB	ATSAW-VSA	ATSAW-ITP	ATSAW-Spacing	Airborne separation (ASEP)	Self-separation (SSEP)	ATSAW-SURF and ADS-B APT	ATS surveillance
Identification								
Callsign				✓	✓	✓	✓	✓
Address	✓	✓	✓	✓	✓	✓	✓	✓
Category					✓	✓	✓	✓
Mode A code								✓
State vector								
Horizontal position	✓	✓	✓	✓	✓	✓	✓	✓
Vertical position	✓	✓	✓	✓	✓	✓		✓
Horizontal velocity	✓	✓	✓	✓	✓	✓		✓
Vertical velocity	✓	✓	✓	✓	✓	✓		✓
Surface heading							✓	
Ground speed							✓	
Mode status								
Emergency/priority status	✓	✓	✓	✓				✓
Capability codes				✓	✓	✓	✓	✓
Operational modes				✓	✓	✓	✓	✓
State vector quality indicator								
NIC	✓	✓	✓	✓	✓	✓	✓	✓
NACp	✓	✓	✓	✓	✓	✓	✓	✓
NACv		✓	✓	✓	✓	✓	✓	✓
SIL		✓	✓	✓	✓	✓	✓	✓
SDA	✓	✓	✓	✓	✓	✓	✓	✓
Air-reference vector						✓		
Intent data								

Table 5.2 Minimum required ADS-B performance for airborne surveillance applications

Performance metric	Required ADS-B performance				IM/spacing (en-route/terminal)	Airborne separation (ASEP) (en-route/terminal)
	Situational awareness applications (ATSAW)					
	AIRB	VSA	SURF	ITP		
Accuracy (NAC$_P$)	5	6	7/9[1]	5	6/7	9
Integrity (NIC)	N/A[2]	6	N/A	5	5/7	9
Velocity Accuracy (NAC$_V$)	1	1	2	1	1/2	3
Source Integrity Level (SIL)	N/A	1	N/A	2	2	2
System Design Assurance (SDA)	1	1	1/2[3]	2	$< 1 \times 10^{-6}$/flight hour	TBD
Update rate (seconds)	3	N/A	≤ 2	≤ 5 to ≤ 24	*[4]	TBD
Latency (seconds)	<1.5	<1.6	<0.5 (on-board)	≤ 4.575	*[4]	TBD

Notes

1 SURF surface targets require NAC$_P$ >= 9. SURF airborne targets require NAC$_P$ = 7 or 9 depending on parallel runway spacing.

2 N/A – not applicable.

3 Hazard level for own ship when airborne or on surface >80 knots = Major (SDA = 2) Hazard level for own ship when airborne or on surface

   <80 knots = Minor (SDA = 1).

4 Not available at the time of writing.

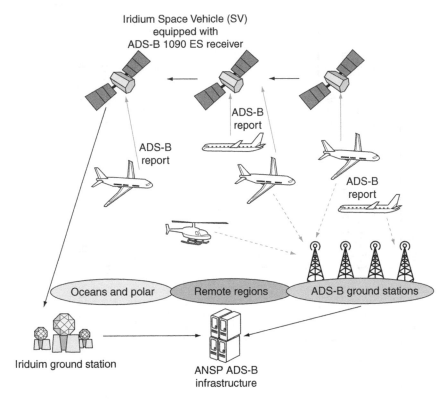

Figure 5.3 Space-based ADS-B system.
Source: Aireon, 2016a.

from Algiers to Marseille, declared a mid-air emergency and dis-appeared from radar display for almost an hour before it finally landed safely back at its point of origin. The investigation concluded that ADS-B Out coverage was limited in the area in which the aircraft "disappeared". Such safety incidents are envisioned to be overcome by the space-based ADS-B system.

In terms of safety improvements, the space-based ADS-B system will allow seamless tracking of the location and position of ADS-B-equipped aircraft globally in real time. Space-based ADS-B system integration into ATM systems can serve as a contingency surveillance layer to limit the impact of a legacy surveillance outage or system failure, or sole-surveillance source in the areas where no surveillance currently exists, or an augmented surveillance to fill in gaps in the existing surveillance coverage.

5.1.1.2.1 SPACE-BASED ADS-B OPERATIONAL PRINCIPLES
AND INFRASTRUCTURES

Figure 5.3 illustrates space-based ADS-B system operations. Principles of the system operation are similar to ADS-B system operations explained in Chapter 4. The additional aspect will be receipt of ADS-B messages broadcast (ASTERIX Cat 21) by the 1090ES receivers installed in the Iridium Space Vehicles (SV). The messages will then be transmitted to the Iridium Ground Stations and finally sent to the ANSP's ADS-B infrastructure for ATC surveillance use.

The system specifications are provided in Table 5.3. The ANSP and airline operators do not require providing any additional infrastructure to support the space-based ADS-B system, in addition to the existing aircraft equipage and ADS-B ground stations required for ADS-B Out operations. The system supports equipage version DO-260, DO-260A and DO-260B. It also uses the top-mounted antenna on the aircraft, currently being used for TCAS.

5.2 How ADS-B can improve the limitations in the radar system?

It is clear that the performance of ADS-B is such that it has the potential to support the increase in air travel demand which current surveillance systems, given their operational limitations, cannot achieve. Table 5.4 shows a comparison of the ADS-B system and the radar systems based on the surveillance applications required by ICAO to accommodate the increasing air travel demand.

Table 5.3 Space-based ADS-B system specifications

Specifications	
Avionics	1090ES ADS-B (DO-260 versions 0, 1, 2)
Range	1100 NM
System coverage	Continuous Global Coverage
Availability	≥99.9%
Latency	≤1.5 seconds to the ATM Automation Platform
Update interval	<8 seconds
Possible separation	Oceanic: ≤15 NM Terrestrial en-route: 5 NM

Source: Aireon, 2016b.

Table 5.4 A comparison of the ADS–B system and the radar system based on applications required by ICAO to accommodate increasing traffic

Surveillance application	Radar systems	ADS-B system
Improved situational awareness of the air traffic	Limited situational awareness in remote and oceanic areas, restricted to line of sight and subject to severe signal fading and interference.	Situational awareness in a particular area depends on the movement of ADS–B equipped aircraft in that area at time (t). Aircraft are independent of ATC to obtain situational awareness. Seamless situational awareness for ATCo with space-based ADS-B.
Radar equivalent	Able to detect both cooperative (SSR) and non-cooperative (PSR) target within limited range.	Provides improved accuracy and higher position update rate, which will enhance the surveillance service. Only detects ADS-B equipped targets.
Enhanced visual acquisition	Limited visual acquisition. Conventional "see and avoid" has reached its limit due to the increasing speed of aircraft, the poor visibility in modern cockpit and flight crew workload in some phases of flight.	Provides enhanced visual acquisition capability with respect to "see and avoid" procedure which applies to VFR/VFR and IFR/VFR operations. This is provided by use of Cockpit Display of Traffic Information (CDTI) application.

Airport surface operations	The Surface Movement Radar (SMR) output deteriorates during heavy rain and vanishes due to line of sight.	ADS-B provides a new source of airport surveillance information for safer and more efficient ground movement management in airports. Airport ground vehicles should also be equipped with ADS-B to generate a complete situational awareness display.
Enhanced separation	Provide low update rate of 4–12 seconds.	The improved accuracy and an update rate of 1–2 seconds enable reduced separation. Subsequently, this will enable redistribution of tasks related to sequencing and merging of traffic between the ATC and aircraft. It will also enable In-Trail Procedures in non-radar airspace, allowing ADS-B equipped aircraft to descend and climb through each others' flight level. This will result in optimized airspace capacity utilization.

Table 5.5 Potential mitigations for the limitations of the radar system

Derived causal factors for current system limitations (Chapter 3)	Potential mitigation based on ADS-B system
C1 Lack of situational awareness	Flight crew will have complete situational awareness of all ADS-B equipped aircraft and geographical structure for a specified range. Controllers will also have complete visual representation of air traffic in their territory.
C2 Limited surveillance coverage	ADS-B provides surveillance coverage in both radar and non-radar airspace.
C3 Inaccurate positioning information	ADS-B provides accurate positioning information derived from GNSS.
C4 Low update rate (position data)	ADS-B provides an update rate of 1–2 seconds in en-route and less than 1 second update rate in the terminal area.
C5 Loss of communication	ADS-B provides continuous broadcast of the aircraft position to ATC centres on the ground.
C6 Unsynchronized surveillance information between flight crew and ATC	Flight crew and controllers will have similar level of situational awareness.
C7 Visual deficiencies in extreme weather conditions	Cockpit Display of Traffic Information (CDTI) application supported by the ADS-B system will provide complete visual aid to flight crew despite weather conditions.

Potential mitigations with ADS-B system for the causal factors due to the limitations in the radar systems (derived in Chapter 3) are described in Table 5.5.

5.3 Summary

This chapter has identified and validated that, limitations exist in the current surveillance systems. The impacts of these limitations are increasing incidents and inability to support the required applications to cater for enhanced flight operations and future air traffic volume. In order to solve these issues, the capabilities of the ADS-B system are analysed. Findings in section 5.1 and analysis in Tables 5.4 and 5.5 show that the system has promising capabilities to improve airspace capacity by providing accurate aircraft positioning information at high update rates and enhance traffic handling capacity by providing separation assistance and situational awareness to flight crew and ATCo. In addition, high system performance (i.e. accuracy, reliability, integrity, availability) and seamless coverage in comparison to the radar systems should lead to improved system safety. However, the safety level of the ADS-B system has to be assessed and validated in various operational environments prior to consideration for global implementation by all stakeholders.

6   ADS-B security

6.1 Introduction

Advancements in information technology, computing infrastructures and communication network technologies have generally made human life easier. However, the ease doesn't always bring goodness. It may also open doors to threats from those with negative motives. An example of threat is cyber-attack. Sole dependency on the technologies does come with a price. The threats may apply to any sectors (banking, communication, media, education, entertainment, transportation, etc.).

ICAO has stressed including provisions for the protection of critical information and communication technology systems against cyber-attacks and interference as stated in the Aviation Security Manual Document 8973/8. This has been further emphasized in the Air Traffic Management Security Manual Document 9985 AN/492 to protect Air Traffic Management systems against cyber-attacks.

This chapter focuses on the ADS-B system security. Before we dig deeper into the topic, the definition of the term "security" and the relationship between "security" and "safety" has to be clear. The *Oxford Dictionary* defines security as "a state of being free from danger and threat". A system is said to be insecure if the system has known vulnerabilities that can be exploited to cause safety impacts. The threats as a result of exploiting the vulnerabilities within the system may lead to safety occurrences. In the aviation environment, the Global Air Traffic Management Operational Concept (ICAO, 2005) defines the security expectation of an integrated, interoperable and globally harmonized ATM system as:

> the protection against threats that stem from intentional acts (e.g. terrorism) or unintentional acts (e.g. human error, natural disaster) affecting aircraft, people or installations on the ground.

Adequate security is a major expectation of the ATM community and of citizens. The ATM system should therefore contribute to security, and the ATM system, as well as ATM-related information, should be protected against security threats. Security risk management should balance the needs of the members of the ATM community that require access to the system, with the need to protect the ATM system. In the event of threats to aircraft or threats using aircraft, ATM shall provide the authorities responsible with appropriate assistance and information.

McCallie *et al.* (2011) add that security assessments analyse the ability of a system to maintain confidentiality, integrity and availability. This definition summarizes the former definition in the Global Air Traffic Management Operational Concept (ICAO, 2005). This chapter will adopt this definition to probe into ADS-B security.

6.2 Vulnerabilities in the ADS-B system

Vulnerability in this section refers to certain characteristics of the system that make it susceptible to a threat or various types of threats. Despite its potential to mitigate the limitations in the radar systems as discussed in Chapter 5, the ADS-B system is still vulnerable to security threats.

6.2.1 Broadcast nature of RF communication

ADS-B's principle of operation, system components, integration and operational environment are discussed in detail in Chapter 4. The ADS-B system broadcasts ADS-B messages containing aircraft state vector information and identity information via RF communication links such as 1090ES, UAT or VDL Mode 4. The broadcast nature of these wireless networks without additional security measures is the main vulnerability in the system. According to Strohmeier *et al.* (2015), contrary to wired networks, there are no challenging obstacles to access a wireless network. Hence it can be easily attacked by any malicious parties.

6.2.2 No cryptographic mechanism

Cryptographic mechanism refers to encryption and decryption of data for confidentiality reasons. Neither the ADS-B messages are encrypted by the sender at the point of origin, nor are the transmission data links. The ICAO (Airports Authority of India 2014) has verified that there is no cryptographic mechanism implemented in the ADS-B protocol.

6.2.3 ADS-B commercial off-the-shelf (COTS) equipment/software

ADS-B equipment, for example receivers, is available as commercial off-the-shelf (COTS) equipment at an affordable price. The receiver can be used to track ADS-B capable aircraft flying within a specific range of the receiver. Furthermore, in 2010, a mobile application called Plane Finder AR was released, enabling precise tracking of aircraft using ADS-B transmissions (NDTV, 2010). Recently, ADS-B receivers have turned into unquestionably the hottest iPad accessory. These all-in-one devices stream GPS, subscription-free weather, traffic and sometimes even backup attitude information to the iPad, changing it from a static chart viewer to an interactive in-flight tool. The receivers work with various apps such as ForeFlight, Garmin Pilot, WingX and FlyQ (Sporty's Pilot Shop, 2015). This clearly indicates that anyone with the available gadgets can track ADS-B equipped aircraft flying in the airspace.

6.2.4 Shared data

As a result of commercially available ADS-B receivers, various parties, either private or public, are sharing real-time air traffic information on the internet. This trend emerged due to the current and next generation ATM systems like NextGen (Federal Aviation Administration, 2012) and SESAR (SESAR, 2008), which demand more information sharing among a wide spectrum of aviation stakeholders (Civil Air Navigation Service Organization, 2014). There are a number of websites on the internet that provide digitized live ADS-B traffic data to the public, e.g. flightradar24.com, radarvirtuel.com and flightaware. com. The availability of the data and the capability to track individual aircraft movements open the door to malicious parties to perform undesired acts that may have safety implications.

6.2.5 ASTERIX data format

All-purpose Structured EUROCONTROL Surveillance Information eXchange (ASTERIX) is a binary format for information exchange in aviation. The ASTERIX standard specification is developed and endorsed by EUROCONTROL. (The ASTERIX specification documents are available at EUROCONTROL, 2016a.) ADS-B data is encoded into ASTERIX CAT-21 format and transmitted by ADS-B equipped aircraft to ADS-B ground stations, and decoded into useable form for ATC use. The ASTERIX format decoding guidance, source codes and tools are widely available in the public domain.

6.2.6 Dependency on on-board transponder

As explained in Chapter 4, ADS-B message encoding and broadcast are performed by either the transponder (for 1090ES) or an emitter (for UAT/VDL Mode 4) on board the aircraft. This indicates that ADS-B based aircraft surveillance is dependent on the on-board equipment. Unfortunately, there is vulnerability in this context, whereby the transponder or emitter on board can be switched off intentionally or unintentionally by anyone in the cockpit. Once it is switched off, ADS-B message broadcast will halt, causing the ATC on the ground or other ADS-B In equipped aircraft in the surrounding area unable to track the particular aircraft. This act will also affect aircraft surveillance via SSR that also relies on the transponder. In addition, the TCAS operation integrity will be affected, due to invisibility of the aircraft.

6.2.7 Complex system architecture

ADS-B is an integrated system, dependent on an on-board navigation system to obtain information about the state of aircraft as well as a communication data link to broadcast the information to ATC on the ground and other ADS-B equipped aircraft. In addition, the system also interacts with external elements such as human (controllers and pilots) and environmental factors. The integrated nature of the system increases the system's vulnerability. For example, the vulnerabilities of the GNSS on which the system relies to obtain aircraft positioning information, are inherited by the system. (Vulnerabilities of the GNSS system are provided in Royal Academy of Engineering, 2011.) Similarly, vulnerabilities of the communication links are also inherited by the ADS-B system.

6.3 Potential threats

In the context of this book, a threat is defined as an action exploiting vulnerabilities of the system to cause damage or harm specifically to the aircraft and generally to the Air Traffic Services (ATS), intentionally or unintentionally. This section discusses the potential threats from malicious parties exploiting the vulnerabilities in the ADS-B system described the previous section. Further, the impact of the potential threats will be analysed in terms of the system's confidentiality, integrity and availability.

6.3.1 Eavesdropping

The broadcast nature of ADS-B RF communication links without additional security measures (cryptographic mechanism) enables the act of eavesdropping into the transmission. This act is also known as aircraft reconnaissance (McCallie *et al.*, 2011). Many organizations or free services use this aspect of the system to provide air traffic information on the internet, e.g. flightradar24.com, radarvirtuel.com and flightaware.com. However, eavesdropping can lead to serious threats such as targeting specific aircraft movement information with the intention to harm the aircraft. This could be done with more sophisticated traffic and signal analyses using available sources such as a Mode S and ADS-B capable open-source GNU Radio module (Airports Authority of India, 2014). This potential threat proves that the system is incapable of maintaining the system's confidentiality and hence affects the system security.

6.3.2 Data-link jamming

Data-link jamming is an act of deliberately/non-deliberately blocking, jamming or causing interference in wireless communication. Deliberate jamming, using a radio jammer device, aims to disrupt information flow (message sending/receiving) between the users within the wireless network. Jammer devices can be easily obtained as COTS devices. According to Strohmeier *et al.* (2015) jamming is a common problem for all wireless communication. However, the impact is severe in aviation due to the system's large coverage area (airspace), which is impossible to control. Furthermore, it involves safety critical data. ADS-B system jamming stops aircraft or ground stations or multiple system users within a specific area from being able to send or receive messages. This threat impacts the ADS-B system's security in terms of its availability.

Deliberate ADS-B system jamming occurs when a malicious party uses the jammer device to block signals on ADS-B frequencies e.g.

1090 MHz. Targeted jamming attack would disable ATS at any airports or Air Traffic Control Centres (ATCC). Jamming a moving aircraft is feasible, but considered to be more difficult (Strohmeier *et al.*, 2015).

ADS-B system transmitting on 1090ES is prone to unintentional signal jamming due to the use of the same frequency (Mode S 1090 MHz) by many systems such as SSR, TCAS, MLAT and ADS-B, particularly in dense airspace. In addition, Ali *et al.* (2014) identified that jamming, deliberately or otherwise, of GPS transmissions from a satellite will also affect the ADS-B system. Not only ADS-B is prone to jamming, so is the SSR system. However, it is more difficult to jam a PSR system due to its rotating antenna and higher transmission power. (Further information on radar jamming is provided in Adamy, 2001.)

6.3.2.1 Types of jamming threats for ADS-B

Apart of GNSS (positioning source for ADS-B) jamming, the main jamming threats for the ADS-B system include Ground Station Flood Denial and Aircraft Flood Denial (McCallie *et al.*, 2011).

6.3.2.1.1 GROUND STATION FLOOD DENIAL

The ground station flood denial blocks 1090 MHz transmissions at the ADS-B ground station. There is no difficulty in gaining close proximity to a ground station. Therefore jamming can be performed using a low-power jamming device to block ADS-B signals from aircraft to the ground station. This threat does not target individual aircraft. It blocks ADS-B signals from all aircraft within the range of the ground station.

6.3.2.1.2 AIRCRAFT FLOOD DENIAL

Aircraft flood denial blocks signal transmission to an aircraft. This threat disables the reception of ADS-B In messages, TCAS and interrogation signals from WAM/MLAT and SSR. It is very difficult to gain close proximity to a moving aircraft. Thus, the attacker needs to use a high-power jamming device. According to McCallie *et al.* (2011), high-power jamming devices are not easily obtained. Furthermore, the jamming function will be ineffective as soon as the aircraft moves out of the specific range of the jamming device. Alternatively, better attempts can be made from within the aircraft. However, this is not an easy task due to the large size of the jamming device. Yet, future technology may be able to reduce the size of the jamming device to pocket size. Hence, the most practical way to attempt the threat is from close proximity to the airport, targeting aircraft that are taking-off, landing or taxiing.

6.3.3 Signal spoofing

ADS-B signal spoofing attempts to deceive an ADS-B receiver by broadcasting fake ADS-B signals, structured to resemble a set of normal ADS-B signals, or by re-broadcasting genuine signals captured else-where or at a different time. Spoofing is also known as message injection (McCallie *et al.*, 2011; Strohmeier *et al.*, 2015). In simpler terms, fake (ghost) aircraft are introduced into the air traffic.

The vulnerability of the system – having no authentication measures implemented at the system's data link layer – easily enables this threat (Strohmeier *et al.*, 2015). The issue of ADS-B spoofing was raised in 2006 when the ability of malicious parties to generate as many as 50 false targets on the radar screens of air traffic controllers was discovered (Wood, 2006). According to Dick Smith, former chairman of Australia's Civil Aviation Administration, this is possible by just using a general aviation transponder, a computer and a $5 antenna to create and inject a 112-bit message that conforms to ADS-B messaging protocol (McCallie *et al.*, 2011).

The impact of this threat is on the integrity of ADS-B based air traffic information, which could lead to undesired operational decisions not only by the controllers but also by airborne and ground surveillance applications (described in Chapter 5). The threat evolved as a result of the emergence of the ADS-B In and Out systems.

6.3.3.1 Types of spoofing threats for ADS-B

Two types of spoofing can be attempted to harm the ADS-B system and subsequently cause safety implications to ATC operations and/or aircraft: Ground Station Target Ghost Injection/Flooding and Aircraft Target Ghost Injection/Flooding.

6.3.3.1.1 GROUND STATION TARGET GHOST INJECTION/FLOODING

Ground station target ghost injection/flooding is performed by injecting ADS-B signals from a single aircraft or multiple fake (ghost) aircraft into a ground station. This will cause the single/multiple fake (ghost) aircraft to appear on the controller's working position (radar screen) (McCallie *et al.*, 2011).

6.3.3.1.2 AIRCRAFT TARGET GHOST INJECTION/FLOODING

The concept of aircraft target ghost injection/flooding is similar to ground station target ghost injection/flooding, except the fake ADS-B signals are injected to an aircraft. Similarly the signals can be a single/multiple fake (ghost) aircraft. This will cause the single/multiple fake (ghost) aircraft to appear on the TCAS and CDTI screens in the cockpit. Making it worse, the fake data will also be used by airborne applications such as ACAS, ATSAW, ITP and others (refer to Chapter 5) for aiding aircraft navigation operations.

Unfortunately it is almost impossible to detect this threat on an aircraft compared to the ground station. This is due to the availability of correlation mechanisms performed on the ATC systems, in particular the Display and Processing of Surveillance Data system that utilizes surveillance data from various sources including PSR and Flight Plan. Fortunately, it would be very difficult to attempt to spoof fake ADS-B signals into a moving aircraft at high altitudes. However, it would be simple to perform the threat on a ground station.

6.3.4 ADS-B message deletion

An aircraft can be made to look like it has vanished from the ADS-B based air traffic by deleting ADS-B message broadcast from the aircraft. This can be done using two methods: destructive interference and constructive interference.

6.3.4.1 Destructive interference

Destructive interference is performed by transmitting an inverse of an actual ADS-B signal to an ADS-B receiver (ground station or ADS-B In). Figure 6.1 illustrates this method. The two signal waves (original and actual) have to be of the same frequency and travelling in the same direction. When the first signal wave is up, the second signal wave is down (inverse) and the two add up to zero. When the first signal wave is down and the second is up (again inverse), they again add up to zero. In reality, the two opposite signal waves cancel each other out and hence no signal wave is left.

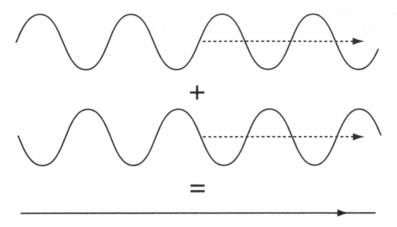

Figure 6.1 Destructive interference.
Source: Gibson, 2016.

In order to perform this threat, an attacker needs to capture the actual ADS-B signal from the target aircraft and inverse the signal and then re-transmit the signal. However, this method requires very precise and complex timing requirements, making it extremely difficult to implement (Strohmeier *et al.*, 2015).

6.3.4.2 *Constructive interference*

Constructive interference is performed by transmitting a duplicate of an ADS-B signal and adding the two signal waves (original and duplicate). The two signal waves (original and actual) have to be of the same frequency and travelling in the same direction. Figure 6.2 illustrates this method. The resultant signal wave will look like the original wave but with a larger amplitude. As a result, the signal will be discarded by the ADS-B receiver due to the CRC algorithm rules in the receiver that only corrects a maximum of five bit errors per message. If the message exceeds this threshold, the message will be assumed as corrupted and discarded. This method does not have timing constraints.

6.3.5 *ADS-B message modification*

ADS-B message modification is feasible on the physical layer during transmission via the data-links using two methods: signal overshadowing and bit-flipping (Strohmeier *et al.*, 2015).

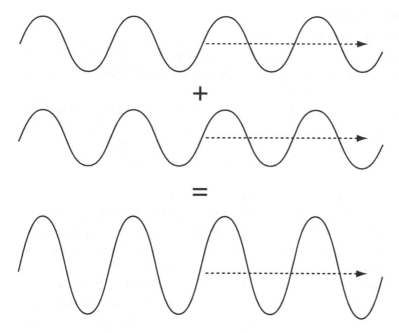

Figure 6.2 Constructive interference.
Source: Gibson, 2016.

6.3.5.1 Signal overshadowing

Signal overshadowing is done by sending a stronger signal to the ADS-B receiver, whereby only the stronger of two colliding signals is received. This method will replace either the whole target message or part of it.

6.3.5.2 Bit-flipping

Bit-flipping is an algorithmic manipulation of binary digits (bits). The attacker will change bits from 1 to 0 or vice versa. This will modify the original ADS-B message.

The attacker's goal is to violate the integrity of ADS-B messages, for example by modifying the aircraft identification (callsign, 24-bit ICAO address) or the aircraft trajectory, tricking the ADS-B receiver (ADS-B In or ground station) to accept a message of the attacker's choice, while the actual sender (aircraft) considers its original message to be delivered successfully. This threat will disrupt ATC operations or aircraft navigation.

6.3.6 Halt ADS-B

Halting the ADS-B transmission can be intentional or unintentional. An unintentional halt can occur to disruption of the GNSS service. As we are aware, the ADS-B system is dependent on GNSS to obtain aircraft positioning information.

An intentional halt is mainly due to the hijacking of an aircraft. The first step to attempt a crime is to make the crime scene invisible. For example, to conduct a robbery in a secured building, the robber would block all the CCTVs at the building. Similarly, hijackers of an aircraft would make themselves invisible.

At present, there are many layers of aircraft surveillance in most of the airspace globally including PSR, SSR, WAM, ADS-B and ADS-C. All of these technologies, apart from PSR, constitute cooperative surveillance systems, requiring an active transponder on board the aircraft. Otherwise the systems are useless. If the controllers in a specific airspace are relying only on either one or all of these surveillance systems for ATC operations, the systems will not be able to provide information (identification and state vector) of an aircraft flying in their airspace without an active transponder on board the aircraft.

Therefore, hijackers on an aircraft would switch off the on-board transponder, to make the aircraft invisible on the controller's working position (radar screen). This action will halt transmission of ADS-B messages, ADS-C transmissions and replies to SSR interrogations. This threat affects the ADS-B systems availability and hence security.

6.3.6.1 Related case study: MH370 incident

This vulnerability was an obvious security threat, in the case of the MH370 incident.

> Malaysia Airlines Flight 370 (MH370) was a scheduled international passenger flight operated by Malaysia Airlines that disappeared on 8 March 2014 while flying from Kuala Lumpur International Airport, Malaysia, to Beijing Capital International Airport in China. The aircraft last made voice contact with air traffic control at 01:19 MYT, 8 March (17:19 UTC, 7 March) when it was over the South China Sea, less than 40 minutes after takeoff. It disappeared from air traffic controllers' radar screens at 01:22 MYT. Malaysian military radar continued to track the aircraft as it deviated westwards from its planned flight path and

crossed the Malaysian Peninsular. It left the range of Malaysian military radar at 02:22 while over the Andaman Sea, 200 nautical miles (370 km) north-west of Penang in north-western Malaysia. The aircraft, a Boeing 777–200ER, was carrying 12 Malaysian crew members and 227 passengers from 15 nations.

(Malaysian ICAO Annex 13 Safety Investigation Team for MH370, 2016)

The report indicates that the aircraft was still detected by the military radar (PSR), until 02:22 MYT. This was after the aircraft disappeared from the air traffic controllers' radar screens at 01:22 MYT. This indicates the possibility of a hijacking that involved the deactivation of the on-board transponder by a malicious party, causing the ADS-B to halt and disabling the SSR/WAM, rendering it impossible to detect the aircraft.

6.4 Security recommendations

This section discusses security recommendations for the ADS-B system. From the introduction of the ADS-B system, no requirements for system security were proposed or required by the system stakeholders. None of the existing standards or minimum operational requirements (RTCA, 2006a, 2011b; ICAO, 1990) by ICAO and RTCA highlights security requirements for ADS-B. Australia pioneered ADS-B system implementation in their airspace and after more than ten years now, Australian airspace is fully covered by ADS-B. Even after such a long duration, the concern regarding ADS-B security has only recently been globally highlighted.

The concern of ADS-B security was overlooked due to the stakeholders' focus on providing surveillance coverage in non-radar airspace. However, recent aircraft incidents such as the September 11 attacks in the United States, Germanwings 4U 9525, MH370 and many other security related incidents have made the aviation industry realize the lack in security is becoming crucial. According to Strohmeier *et al.* (2015) the security issue in the ADS-B system has gained wide media coverage due to two talks presented at the DEFCON (Kunkel, 2009) and Black Hat (Costin and Francillon, 2012) security conferences.

McCallie *et al.* (2011) state that in order to increase ADS-B security, its functionalities need to be reduced. This is somewhat true. For example, one of the main objectives of ADS-B is to provide a similar level of situational awareness to all users via data sharing. However, if the broadcast channel is encrypted, the objective would be more difficult to achieve. Inline with this statement, the researchers argue further that, both increasing and decreasing the security level of ADS-B will have a negative impact on safety. Therefore, trade-offs need to be identified, assessed and analysed carefully to ensure secure and safe ATC operations.

6.4.1 ADS-B security requirements

In order to derive ADS-B security requirements, we need to go back to the definition discussed earlier in section 6.1. The security requirements should enable the system to maintain its confidentiality, integrity and availability. Strohmeier *et al.* (2015) outlined a set of security requirements for the ADS-B system in line with the definition:

- *Data integrity*
  The system security should be able to ensure that the ADS-B data received by the ground station or other aircraft (ADS-B In) are the exact message transmitted by the aircraft. In addition, it should also be able to detect any malicious modification to the data during the broadcast.
- *Source integrity*
  The system security should be able to verify that an ADS-B message received is sent by the actual owner (correct aircraft) of the message.
- *Data origin (location) authentication*
  The system security should be able to verify that the positioning information in the ADS-B message received is the original position of the aircraft at the time of transmission.
- *Low impact on current operations*
  The system security hardware/software should be compatible with the current ADS-B installations and standards.
- *Sufficiently quick and correct detection of incidents*
- *Secure against DoS-attacks against computing power*
- *System security functions need to be scalable irrespective of traffic density*
- *Robustness to packet loss*
  For example, a jammed wireless channel should decrease neither the security level nor the reliability of the system.
- *Non-repudiation*
  Achieving non-repudiation is seen as nice to have but not very high on the priority list for immediate air traffic security and is more of a legal topic.

6.4.2 Potential mitigations for ADS-B vulnerabilities

Vulnerabilities in the ADS-B system are discussed in detail in section 6.2. At this point, we should be aware that the threats (section 6.3) that may cause safety impacts are due to exploitation of the vulnerabilities in the system. Therefore, the solution should be looking at efforts to eliminate or at least reduce the level of vulnerabilities within the ADS-B system. This section will try to present potential solutions, where possible, for the vulnerabilities identified in section 6.2.

The implementation of broadcast authentication using a secure and effective method will resolve the vulnerabilities of the system's open nature and unavailability of cryptographic mechanisms, and threats (message spoofing, jamming and modification) using the easily available COTS can be eliminated. However, the objective of ADS-B to provide data sharing can still be achieved in a secured environment. A number of methods for the broadcast authentication exist and are implemented in various other industries. These include Non-Cryptographic Schemes on the Physical Layer, Public Key Cryptography and Retroactive Key Publication (for details on these methods see Strohmeier et al., 2015). The technology of Virtual Private Network (VPN) for wireless networks is also worth looking into. These methods may or may not be suitable to the ADS-B system. Rigorous research, assessment and analysis need to be conducted to choose the best method for the system. Most important is to keep the functionalities of the system as much as possible while maintaining the systems confidentiality, integrity, availability and safety.

ADS-B's dependency on on-board transponders is another vulnerability of the system. In order to resolve this, manual access to the transponder should not be given to the personnel in the cockpit. As soon as the aircraft engine is powered, the transponder should automatically activate and deactivate as soon as the engine stops.

ADS-B will inherit the vulnerabilities of GNSS, the source of its positioning information. To detect the impact of the vulnerabilities on the positioning information received on board from the GNSS, an alternative navigation source, e.g. INS/IRS, can be used for verification purposes. To conduct the verification at the ground station, a number of methods can be applied: Multilateration, Distance Bonding, Kalman Filtering, Group Verification, Data Fusion or Traffic Modelling (Strohmeier et al., 2015). Again, the accuracy and effectives of these methods need to be validated with rigorous efforts.

Conclusion

Increasing air traffic

The ability to meet increasing air travel demand is determined by a number of capacity drivers: the controller's capacity, tools to aid the controller's task, the extent of the pilot's situational awareness, aircraft capability and contextual environment including the weather conditions and airport capacity. In a structural interview with Harry Daily from the Civil Aviation Authority (CAA), he argued that the ability of the ANSP to provide the appropriate level of air navigation services is also one of the critical factors in meeting the future air travel demand. The controller's workload and task time directly impacts the number of aircraft that can be managed by them in their airspace. Hence, airspace capacity utilization depends on the capacity of the controllers to manage the airspace safely. However, the controller's workload can be improved by having reliable and accurate tools to assist the controller's tasks.

In addition, shared surveillance information between the controllers and pilots can enhance the situational awareness and reduce the procedural control. As a result, there will be less communication between the pilot and controller. The enhanced situational awareness of the pilot also enables more flexible flight routings especially in oceanic airspace. This will reduce the flight time and enable a more economical flight by saving fuel. In addition, a more complete situational awareness on board will also eliminate difficulties in navigation during extreme weather conditions.

The ability of the surveillance and navigation systems to provide accurate aircraft position information, intent data, speed and the higher update rates will permit the setting of reduced separation minima. Consequently, airspace utilization can be optimized. In order to support the advanced systems and technologies, aircraft and airports have to be equipped with necessary equipment and infrastructure. As a result, a safe and efficient gate-to-gate operation can be achieved and the increasing air travel demand can be met.

In order to achieve these, are the radars that we have at present sufficient? Can ADS-B resolve the problems?

Radar and ADS-B

Chapter 3 has identified the limitations in the radar systems. The limitations include unavailability of services in oceanic and remote areas, limited services during extreme weather conditions and outdated equipment with limited availability of spare parts to support system operation and low performance in terms of accuracy, integrity, continuity and availability, particularly in high-density traffic areas including airports. The impacts of these limitations are increasing incidents and an inability to support the required surveillance applications for enhanced flight operations and future air traffic volume.

ADS-B was proposed by the ICAO and is envisioned to fill the gaps in the radar systems. In line with this, the FAA's NextGen and EUROCONTROL's SESAR programmes recognize ADS-B as the key to their respective goals to modernize the ATM operations and address the limitations in the radar systems. Therefore the capabilities of the ADS-B system were analysed. Findings show that the system has promising capabilities to improve airspace capacity by providing accurate aircraft positioning information at high update rates and enhance traffic handling capacity by providing separation assistance and situational awareness to flight crew. In addition, high system performance (i.e. accuracy, reliability, integrity, availability) in comparison to the radar systems should lead to improved system safety.

The improvements include improved situational awareness for pilots in the cockpit despite weather conditions via ADS-B In (CDTI). It also enables a similar level of situational awareness between pilot and controllers. Furthermore, high-performance aircraft state vector information from the ADS-B system has motivated the development of various new airborne applications such as ATSAW-ITP, ATSAW-VSA, ATSAW-SUR, Spacing/Interval-Management (IM), ASEP and SSEP to enhance aircraft navigation on the airport surface, en-route TMA, oceanic and remote airspace with improved safety as discussed in Chapter 5.

However, the ADS-B system has its drawbacks. ADS-B Out's range to the ground stations from oceanic airspace is limited. This drawback has been rectified by the recent extension of the ADS-B Out system to space-based ADS-B. In addition, ADS-B not only relies on an on-board transponder, but also on the integrity of the GNSS or on-board navigation system. If the navigation system fails, the aircraft will not be able to broadcast its position, or worse, it may broadcast invalid positions. ADS-B's high dependency on the navigation and communication systems increases the system's complexity and hence failure modes.

Security

ADS-B and radar (SSR modes A, C and S) have certain common security characteristics, although they don't necessarily share the same susceptibility to exploitation. Furthermore, the vulnerabilities identified in the ADS-B system differ for radar. For example, SSR responds when interrogated by the ground station while ADS-B continually broadcasts aircraft information.

However, FAA claims that both ADS-B and SSR transmissions are susceptible to the intentional introduction of a false target. This claim is acceptable as it is a potential security threat to any unsecured RF communication medium. Both technologies are using the same data link frequency 1090/1030 MHz without any cryptographic mechanisms.

FAA also argued that the impact of spoofing and jamming on the ADS-B system will not impact ATC operations, with the claim that the ADS-B data will be fused with the PSR data before it is displayed on the radar screen at the controller's working position. This claim assumes that data verification using a PSR target will remove all the discrepancies between a spoofed or jammed ADS-B target. On the contrary, the FAA Final Ruling (Federal Aviation Administration, 2010b) states discontinuation of PSR upon full operational ADS-B in the national airspace. Therefore, the argument is invalid. Moreover, data fusion is not feasible in non-radar airspace. For example, in Australia, ADS-B is the sole surveillance means of ATC in less populated parts of the continent. This scenario also applies to other states such as Alaska.

The main vulnerability of the ADS-B system is its broadcast nature without any security measures, which can be easily exploited to cause various other threats such as through eavesdropping aircraft movement with the intention to harm, message deletion and modification. The system's dependency on the on-board transponder is also considered as a major vulnerability, which is shared by the SSR. This vulnerability can be exploited by aircraft hijackers to make the aircraft movements invisible.

In addition, ADS-B's high dependency on communication and navigation (GNSS) systems causes the system to inherit the vulnerabilities of those systems and hence results in more possible threats. In general, advancements in computing, software and hardware technologies are the major aids to malicious parties who wish to carry out the threats by exploiting the vulnerabilities in the ADS-B system.

ICAO in the 13th ADS-B Study and Implementation Task Force meeting has proposed the formation of a working group to analyse and develop mitigation measures for the various vulnerabilities identified in the ADS-B system. Besides, many researches are being conducted by various institutions to resolve ADS-B security issues. Some of the solutions include ADS-B broadcast encryption and authentication, aircraft positioning data integrity verification mechanisms and ADS-B equipment licensing/authorization.

Bibliography

Adamy, D. L. 2001. *EW 101: A First Course in Electronic Warfare*. Boston, MA: Artech House.

Aeronautical Surveillance Panel (ASP) 2007. Surveillance Manual. Draft 2.0 ed. ASP Working Group Meeting, Montreal.

Aireon 2016a. *Space-Based ADS-B: Unlock Your ATM Potential*. www.enav.it/ enavWebPortalStatic/ATM2016/assets/aireon_brochuredef_1.pdf.

Aireon 2016b. Space Based ADS-B. *In:* International Civil Aviation Organization (ed.) *Twenty-First Meeting on the Improvement of Air Traffic Services over the South Atlantic (SAT/21)*. Lisbon, Portugal: ICAO.

Airports Authority of India 2014. *Security Issues of ADS-B Operations*. International Civil Aviation Organization, ADS-B Study and Implementation Task Force. Hong Kong, China.

AirservicesAustralia 2007. ADS-B and TCAS.

Ale, B. 2009. *Causal Model for Air Transport Safety*. Delft: Risk Centre TU Delft.

Ali, B. S. 2016. System Specifications for Developing an Automatic Dependent Surveillance-Broadcast (ADS-B) Monitoring System. *International Journal of Critical Infrastructure Protection*, 15: 40–46.

Ali, B. S., Majumdar, A., Ochieng, W. Y., Schuster, W. and Chiew, T. K. 2015. A Causal Factors Analysis of Aircraft Incidents Due to Radar Limitations: The Norway Case Study. *Journal of Air Transport Management*, 44–45: 103–109.

Ali, B. S., Ochieng, W., Majumdar, A., Schuster, W. and Chiew, T. K. 2014. ADS-B System Failure Modes and Models. *The Journal of Navigation*, 67: 995–1017.

Ali, B. S., Schuster, W. and Ochieng, W. Y. 2016. Evaluation of the Capability of Automatic Dependent Surveillance Broadcast to Meet the Requirements of Future Airborne Surveillance Applications. *The Journal of Navigation*, 71: 49–66.

Avinor 2011. *Avinor* [Online]. Gardermoen, Norway Avinor AS. www. avinor.no/ [accessed 25 October 2011].

Barmore, B., Penhallegon, W. J., Weitz, L. A., Bone, R. S., Levitt, I., Flores, J. A., Kriegsfeld, D. A. and Johnson, W. C. 2016. Interval Management: Development and Implementation of an Airborne Spacing Concept. AIAA Guidance, Navigation, and Control Conference. San Diego, California, USA.

Bloem, E. A., Blom, H. A. P. and Schaik, F. J. V. 2002. *Advanced Data Fusion for Airport Surveillance*. National Aerospace Laboratory.

Brooker, P. 2002. Future Air Traffic Management: Quantitative Enroute Safety Assessment: Part 1 – Review of Present Methods. *The Journal of Navigation*, 55: 197–211.

Brooker, P. 2004. Radar Inaccuracies and Mid-air Collision Risk: Part 2 – Enroute Radar Separation Minima. *The Journal of Navigation*, 57: 39–51.

Butterworth-Hayes, P. 2012. *The Market for Wide-Angle/Beyond-Airport Surface ADS-B Surveillance*. ATC Global Market Intelligence Reports.

Cascade Operational Focus Group 2008. *Use of ADS-B for Enhanced Application of Own Visual Separation by Flight Crew on Approach (ATSA-VSA)*. EUROCONTROL.

Cascade Operational Focus Group 2009. *Use of ADS-B for Enhanced Traffic Situational Awareness by Flight Crew During Flight Operations – Airborne Surveillance (ATSA-AIRB)*. EUROCONTROL.

Civil Air Navigation Service Organization 2014. *Cyber Security and Risk Assessment Guide*. CANSO.

Comsoft GmbH 2007. *Automatic Dependent Surveillance – Broadcast (ADS-B) Deployment*. www.airport-int.com/article/automatic-dependent-surveillance. html [accessed 2013].

Costin, A. and Francillon, A. 2012. *Ghost Is in the Air (Traffic)*. Las Vegas, NV: Black Hat USA.

Dawson, M. 2004. *Understanding Mode S Technology*. Aeroflex-Chandlers Ford UK.

De Oliveira, I. R., Vismari, L. F., Cugnasca, P. S., Camargo Junior, J. B., Bakker, B. G. H. J. and Blom, H. A. P. 2009. A Case Study of Advanced Airborne Technology Impacting Air Traffic Management. *In*: L. Weigang, A. G. de Barros and I. R. de Oliveira (eds) *Computational Models, Software Engineering and Advanced Technologies in Air Transportation: Next Generation Applications*. Hershey, PA: IGI Global.

EASA 2008. *Acceptable Means of Compliance for ADS-B Non-Radar Airspace*. AMC 20–24.

EUROCAE 2010. *Technical Specification for Wide Area Multilateration (WAM) Systems*. 101/ED-142.

EUROCAE 2011. *Safety Performance and Interoperability Requirements for Flight Deck Interval Management (ASPA-FIM)*. ED-195.

EUROCAE and RTCA 2010. *Safety Performance and Interoperability Requirements for ATSAW during light operations (ATSAW-AIRB)*. ED-164/DO-319.

EUROCONTROL 1997. *Eurocontrol Standard for Radar Surveillance in En-Route Airspace and Major Terminal Areas.* EUROCONTROL.

EUROCONTROL 2001–2013. *VHF Digital Mode 4.* www.eurocontrol.int/services/vhf-digital-mode-4 [accessed 15 February 2013].

EUROCONTROL 2005. *ATM Contribution to Aircraft Accidents/Incidents.* Safety Regulation Commission Document.

EUROCONTROL 2006. *Air Traffic Control (ATC).* www.eurocontrol.int/corporate/public/standard.page/cd-airtraffic_controller.html.

EUROCONTROL 2007. *ADS-B for Dummies – 1090MHz Extended Squitter.* EUROCONTROL.

EUROCONTROL 2008. *The ATM Surveillance Strategy for ECAC.* European Air Traffic Management (EATM).

EUROCONTROL 2009a. *CRISTAL-ITP Project.* EUROCONTROL CASCADE Programme.

EUROCONTROL 2009b. *Reporting and Assessment of Safety Occurrences in ATM.* Safety Regulatory Requirement, 3.0 ESARR2.

EUROCONTROL 2010. *Annual Safety Report – 2010.* EUROCONTROL.

EUROCONTROL 2011a. *Surveillance Performance and Interoperability Implementing Rule (SPI-IR).* Commission of the European Union. http://eur-lex.europa.eu/legal-content/EN/ALL/;jsessionid=QcYRTxbZk87bQks5dZtTwvc5zSqF1gM1bvRN1CyQVpJ1tFFnrnfk!-2081994908?uri=CELEX:32011R1207 [accessed 13 April 2013].

EUROCONTROL 2011b. *Transponder.* https://extranet.eurocontrol.int/http://atmlexicon.eurocontrol.int/en/index.php/Transponder#Definition [accessed 1 November 2011].

EUROCONTROL 2016a. *All-purpose Structured EUROCONTROL Surveillance Information Exchange (ASTERIX).* www.eurocontrol.int/services/asterix [accessed 30 August 2016].

EUROCONTROL 2016b. *European ATM Master Plan.* EUROCONTROL. www.atmmasterplan.eu/data/projects/19074 [accessed 20 May 2016].

Federal Aviation Administration 2010a. *Airworthiness Approval of Automatic Dependent Surveillance – Broadcast (ADS-B) Out Systems.* Advisory Circular AC 20–165.

Federal Aviation Administration 2010b. *Automatic Dependent Surveillance Broadcast (ADS-B) Out Performance Requirements to Support Air Traffic Control (ATC) Service.* Final Rule, 14 CFR Part 91. Federal Registers 75 (103).

Federal Aviation Administration 2012. *NextGen Implementation Plan.* FAA.

German Federal Bureau of Aircraft Accidents Investigation 2004. *Investigation Report.*

Gibson, G. N. 2016. *Constructive and Destructive Interference.* High-Intensity Laser Physics. University of Connecticut.

Hansman, J. R. 1997. Integrated Human Centered System Approach to the Development of Advanced Cockpit and Air Traffic Management

Systems. Air Traffic Management Research and Development Seminar, Sacley, France.

Herrera, I. A., Nordskaga, A. O., Myhreb, G. and Halvorsenb, K. 2009. Aviation Safety and Maintenance under Major Organizational Changes, Investigating Non-existing Accidents. *Accident Analysis & Prevention*, 41: 1155–1163.

Hollnagel, E., Hoc, J. M. and Cacciabue, P. C. 1995. Work with Technology: Some Fundamental Issues. *In: Expertise and Technology: Cognition and Human-Computer Cooperation*. New York: Lawrence Erlbaum Associates.

ICAO 1990. *Automatic Dependent Surveillance*. 226 AN/135.

ICAO 1993. *Human Factors Digest no 7: Investigation of Human Factors in Accidents and Incidents*. 240 AN/144.

ICAO 1998a. *Address by the Director of the Air Navigation Bureau of the International Civil Aviation Organization (ICAO) – Mr. Jack Howell at the Official Opening of the World-wide CNS/ATM Systems Implementation Conference*. Rio de Janeiro. http://legacy.icao.int/icao/en/ro/rio/danb.htm [accessed 19 March 2010].

ICAO 1998b. *Air Traffic Services Planning Manual*. Montreal, Canada: International Civil Aviation Organization (ICAO).

ICAO 1998c. *Manual on Airspace Planning Methodology for the Determination of Separation Minima*. 9689 AN/953.

ICAO 1999a. *Manual of Technical Provisions for the Aeronautical Telecommunication Network*. 9705 AN/956.

ICAO 1999b. *Manual on Required Navigation Performance (RNP)*. 9613 AN/937.

ICAO 1999c. Status of Air-Ground Data Link Standardization (Information Paper). 3rd Meeting of CAR/Sam (ICAO). Buenos Aires, Argentina.

ICAO 2000. *National Plan for CNS/ATM Systems*. 278 AN/164.

ICAO 2001. *Aircraft Accident and Incident Investigation* (Annex 13 to the Convention on International Civil Aviation).

ICAO 2002a. *Global Air Navigation Plan for CNS/ATM Systems*. 2nd ed. 9750 AN/963.

ICAO 2002b. ICAO Annex 10 Amendment 77 – Mode S Extended Squitter SARP.

ICAO 2003a. ADS-B Concept of Use (Working Paper 6 – Appendix). 11th Air Navigation Conference (ICAO). Montreal, Canada.

ICAO 2003b. *Report of the Automatic Dependant Surveillance-Broadcast (ADS-B) Study and Implementation Task Force Meeting*. Brisbane, Australia.

ICAO 2003c. Comparative Analysis of ADS-B Data Links (Information Paper 12). 11th Air Navigation Conference (ICAO). Montreal, Canada: International Civil Aviation Organization.

ICAO 2004a. *Advanced Surface Movement Guidance and Control Systems (A-SMGCS) Manual*. 9830 AN/45.

ICAO 2004b. *Manual of Secondary Surveillance Radar (SSR) Systems*. 9684 AN/951.

ICAO 2004c. *Manual on Mode S Specific Services.* 9688 AN/952.

ICAO 2005. *Global Air Traffic Management Operational Concept.* 9854 AN/458.

ICAO 2006a. *ADREP 2000 Taxonomy.*

ICAO 2006b. *Assessment of ADS-B to Support Air Traffic Services and Guidelines for Implementation.*

ICAO 2006c. *Manual on Required Communication Performance (RCP).* 9869 AN/462.

ICAO 2006d. *Procedures for Air Navigation Services – Aircraft Operations (PAN-OPS).* 5th ed. 8168 OPS/611.

ICAO 2007a. *Air Traffic Management (ATM).* 15th ed. 4444 ATM/501.

ICAO 2007b. *CNS/ATM Systems.* www.icao.int/icao/en/ro/rio/execsum.pdf.

ICAO 2007c. *Guidance Material on Comparison of Surveillance Technologies (GMST).*

ICAO 2008. *Performance Based Navigation (PBN) Manual.* 9613 AN/937.

ICAO 2009. *Manual on the Universal Access Transceiver (UAT).*

ICAO 2010. *Report of the 9th Meeting of ADS-B Study and Implementation Task Force.* Jakarta, Indonesia.

ICAO 2012a. *Difference between D0-260, D0-260A and D0-260B.* ADS-B Study and Implementation Task Force.

ICAO 2012b. *Manual on Airborne Surveillance Applications.* 9994 AN/496.

ICAO 2012c. *Report of the 12th Air Navigation Conference.*

Institute of Air Navigation Services 2003. *Automatic Dependent Surveillance.* EUROCONTROL.

Isaac, A. R. and Ruitenberg, B. 1999. *Air Traffic Control: Human Performance Factors.* Aldershot, Hants: Ashgate.

Joint Planning and Development Office (JPDO)-Air Navigation Services Working Group (ANSWG) 2008. FAA's Surveillance/Positioning Backup Strategy Alternatives Analysis. Next Generation Air Transportation System, Paper No.: 08–003.

Kunkel, R. 2009. Air Traffic Control: Insecurity and ADS-B. Defcon 17. Las Vegas, NV.

Lambert, M., Riera, B. and Martel, G. 1999. Application of Functional Analysis Techniques to Supervisory Systems. *Reliability Engineering and System Safety*, 64: 209–224.

Lissandre, M. 1990. *Maitriser SADT.* Paris: Albert Colin.

McCallie, D., Butts, J. and Mills, R. 2011. Security Analysis of the ADS-B Implementation in the Next Generation Air Transportation System. *International Journal of Critical Infrastructure Protection*, 4: 78–87.

Malaysian ICAO Annex 13 Safety Investigation Team for MH370 2016. 2nd Interim Statement. MH370 Safety Investigation. Kuala Lumpur, Malaysia: Ministry of Transport, Malaysia.

NASA 2000. *Required Communication Performance (RCP).* http://as.nasa.gov/aatt/wspdfs/Oishi.pdf.

NATS 2002. Evaluation of ADS-B at Heathrow for the EUROCONTROL ADS Programme Report.

NDTV 2010. A Phone Application That Threatens Security. *Press Trust of India*, 4 October.

Olivier, B., Gomord, P., Honoré, N., Ostorero, L., Taupin, O. and Philippe, T. 2009. Multi Sensor Data Fusion Architectures for Air Traffic Control Applications. *In:* N. Milisavljevic (ed.) *Sensor and Data Fusion*. In-Tech.

Owusu, K. 2003. Draft Material for Manual on ATC Surveillance Architecture. Surveillance and Conflict Resolution System Panel (SCRSP) Working Group-B.

Perrin, E. and Kirwan, B. 2000. A Systemic Model of ATM Safety: The Integrated Risk Picture. EUROCONTROL ATM Seminar.

Rekkas, C. 2013. Progress of WAM, ADS-B Out and ATSAW deployment in Europe. *In:* German Institute of Navigation (ed.) *International Symposium on Enhanced Solutions for Aircraft and Vehicle Surveillance Applications (ESAVS)*. Berlin, Germany.

Royal Academy of Engineering 2011. *Global Navigation Space Systems: Reliance and Vulnerabilities*. London: The Royal Academy of Engineering.

RTCA 1998. Minimum Aviation System Performance Standards for Required Navigation Performance for Area Navigation.

RTCA 2002. Minimum Aviation System Performance Standards for Automatic Dependant Surveillance Broadcast (ADS-B).

RTCA 2006a. Minimum Aviation System Performance Standards for Automatic Dependant Surveillance Broadcast (ADS-B).

RTCA 2006b. Minimum Operational Performance Standards (MOPS) for Traffic Alert and Collision Avoidance System II (TCAS 11) Hybrid Surveillance.

RTCA 2008. Safety, Performance and Interoperability Requirements Document for the In-Trail Procedure in Oceanic Airspace (ATSA-ITP) Application.

RTCA 2011a. Minimum Operational Performance Standards for 1090 MHz Extended Squitter Automatic Dependent Surveillance – Broadcast (ADS-B) and Traffic Information Services – Broadcast (TIS-B). DO-260B.

RTCA 2011b. Minimum Operational Performance Standards for 1090 MHz Extended Squitter Automatic Dependent Surveillance – Broadcast (ADS-B) and Traffic Information Services (TIS-B) with Corrigendum 1.

Selex System Integration 2013. *Transponder*. Christian Wolff. www.radartutorial.eu/13.ssr/sr17.en.html [accessed 16 February 2013].

Sensis Corporation 2009. *An Expanded Description of the CPR Algorithm*. RTCA 1090-WP30–12.

SESAR 2008. The ATM Deployment Sequence D4.

SESAR Joint Undertaking 2011. System Wide Information Management (SWIM). EUROCONTROL and European Commission. www.sesarju.eu/sites/default/files/documents/reports/factsheet-swim.pdf.

Skybrary 2011. *Safety Nets*. www.skybrary.aero/index.php/Safety_Nets [accessed 28 October 2011].

Smith, A., Cassell, R., Breen, T., Hulstrom, R. and Evers, C. 2006. Methods to Provide System-wide ADS-B Back-up, Validation and Security. 25th AIAA/IEEE Digital Avionics Systems Conference.

Sporty's Pilot Shop 2015. *Which ADS-B Receiver Should I Buy?* http://ipad pilotnews.com/2015/10/ads-b-receiver-buy-2/ [accessed 30 August 2016].

Stolzer, A. J., Halford, C. D. and Goglia, J. J. 2008. *Safety Management Systems in Aviation.* Farnham, Surrey: Ashgate.

Strohmeier, M., Vincent, L. and Ivan, M. 2015. On the Security of the Automatic Dependent Surveillance-Broadcast Protocol. *IEEE Communications Surveys & Tutorials*, 17: 1066–1087.

Surveillance and Conflict Resolution Systems Panel 2004a. *Report on Required Surveillance Performance (RSP) for Airborne Surveillance.* SCRSP/1-WP/30.

Surveillance and Conflict Resolution Systems Panel 2004b. *Intercept Functionality.* SCRSP/1-WP/23.

US Air Force 2013. *GPS Satellite Outage Information.* http://adn.agi.com/ SatelliteOutageCalendar/SOFCalendar.aspx [accessed 17 November 2013].

Vidal, L. 2010. ADS-B IN-ATSAW (Airborne Traffic Situational Awareness). CAAC-Thales ADS-B Flight Operation Seminar. Beijing, China.

Vidal, L. 2012. Airborne Traffic Situational Awareness. ICAO ADS-B Study and Implementation Task Force. Jeju: Airbus.

Vismari, L. F. 2007. Vigilancia Dependente Automatica no Controle de Trafego Aereo: avaliacao de risco baseada em Redes de Petri Fuides e Estocasticas. Master's in Engineering Degree Dissertation, School of Engineering, Universaiti of Sao Paulo, Brazil.

Vismari, L. F. and Camargo Jr. J. B. 2005. Evaluation of the Impact of New Technologies on Aeronautical Safety: An Approach through Modelling, Simulation and Comparison with Legacy Systems. *Journal of Brazilian Air Transportation Research Society*, 1: 19–30.

Vismari, L. F. and Camargo Jr. J. B. 2008. An Absolute-Relative Risk Assessment Methodology Approach to Current Safety Critical Systems and Its application to the ADS-B Based Air Traffic Control System. Symposium of Reliable Distributed System. Naples, Italy.

Vismari, L. F. and Camargo Jr. J. B. 2011. A Safety Assessment Methodology Applied to CNS/ATM-based Air Traffic Control System. *Reliability Engineering and System Safety*, 96: 727–738.

Wassan, J. 1994. *Avionics Systems: Operation and Maintenance.* Englewood, CO: Jeppesen.

Wilke, S. and Majumdar, A. 2012. Critical Factors Underlying Airport Surface Accidents and Incidents: A Holistic Taxonomy. *Journal of Airport Management*, 6: 170–190.

Wood, A. 2006. After ADS-B Launch, Security Concerns Raised. *Aviation International News.*

Zwegers, D. 2010. Using ADS-B for Accident Investigation and Prevention. *ISASI Forum*, 43: 23–25.

Index

Page numbers in *italics* denote tables, those in **bold** denote figures.

Printed and bound by CPI Group (UK) Ltd, Croydon, CR0 4YY

01/11/2024

01782621-0020